HOW TO RESTORE YOUR FORD PICKUP

Tom Brownell

MBI Publishing Company

First published in 1993 by MBI Publishing Company, PO Box 1, 729 Prospect Avenue, Osceola, WI 54020-0001 USA

© Tom Brownell, 1993

All rights reserved. With the exception of quoting brief passages for the purposes of review no part of this publication may be reproduced without prior written permission from the Publisher.

The information in this book is true and complete to the best of our knowledge. All recommendations are made without any guarantee on the part of the author or Publisher, who also disclaim any liability incurred in connection with the use of this data or specific details.

We recognize that some words, model names and designations, for example, mentioned herein are the property of the trademark holder. We use them for identification purposes only. This is not an official publication.

MBI Publishing Company books are also available at discounts in bulk quantity for industrial or sales-promotional use. For details write to Special Sales Manager at Motorbooks International Wholesalers & Distributors, 729 Prospect Avenue, PO Box 1, Osceola, WI 54020-0001 USA.

Library of Congress Cataloging-in-Publication Data

Brownell, Tom
How to restore your Ford pickup/Tom Brownell.
p. cm.
Includes index.
ISBN 0-87938-726-2
1. Ford trucks—Conservation and restoration.
I. Title.
TL230.5.F57B76   1993
629.223—dc20   93-7469

On the front cover: A beautifully restored 1950 Ford F-1 owned by Carol and Frank Childs, Boca Raton, Florida. *Mike Mueller*

Printed in the United States of America

Contents

	Acknowledgments	4
Chapter 1	**Introduction**	5
Chapter 2	**Ford Pickup Trucks Through the Years**	8
Chapter 3	**Rebuilding or Restoring**	19
Chapter 4	**Tool Selection**	32
Chapter 5	**Stripping/Derusting**	43
Chapter 6	**Straight Axle Front Ends**	54
Chapter 7	**Engine Overhaul**	67
Chapter 8	**Driveline Overhaul**	89
Chapter 9	**Rebuilding Brakes**	102
Chapter 10	**Metal Repair**	116
Chapter 11	**Priming and Painting**	134
Chapter 12	**Replacing Wiring**	150
Chapter 13	**Redoing the Interior**	166
Chapter 14	**Redoing Rubber**	179
Chapter 15	**Box Restoration**	191
Chapter 16	**Plating**	201
Chapter 17	**Storage**	210
	Appendices	217
	Index	224

Acknowledgments

The following friends and Ford truck authorities assisted in the writing of this book: Jim Baker, Nathan Brownell, Christine Collins of the Eastwood Company, Patrick Dykes, Gary and Crystal Edgerly, Vic Fowler, Noel Glucksman, Bruce Horkey, Tim Master, Dick Matott, Joel Miller, Dave Moyer, Roy Nagel, Herb Parsons, Mike Schmander, Alvin Shier, Jim Simpson, Wayne Tatro of LeBaron Bonney Company, Bob Thatcher, Clarence Thielen, and Herb Ward.

Chapter 1

Introduction

Pickups have eclipsed cars in new-vehicle sales. Could Henry Ford have dreamed this would happen, in 1924 when he bolted a little box in place of the turtle deck of his fast-selling Model T roadsters and created the first "pickup" truck? Ford Motor Company's been a leader in the truck business ever since.

From the start, Ford trucks have been distinctive. The Model T with its planetary transmission made it possible for first-time drivers to sit behind the wheel and take off down the road. Of course, there was also the Model T's other unique characteristic: it came in any color, so long as it was black.

The Model A, which flourishes today in numbers exceeding those of all other makes of its era combined, represented simplicity in perfection. Consider the troublesome "extras" Ford spared Model A owners from having to fix or replace. The car had no fuel pump; gasoline ran to the carburetor from the tank in the cowl by means of gravity feed. The vehicle also had no spark plug wires to break down and deteriorate engine performance. Instead current flowed from the distributor to the spark plugs over thin strips of copper. And original strips, installed on cars at the factory, are still functioning—over sixty years later.

The flathead V-8, introduced in 1932 and built through 1953, gave Ford cars and pickups their performance image. All the speed equipment "goodies" available for these engines in their heyday are still in supply today.

To a large measure, Ford's sales success has been built on putting the buyer first. As early as the Model T, Old Henry insisted on using strong, vanadium steel for the axles and other chassis parts of his cars and trucks. The construction looked spindly, but the steel was the strongest made, and its quality gave Ford a reputation for durability that has become legendary.

When Dearborn's designers were restyling Ford pickups and heavier-duty trucks after World War II, they remembered to keep the buyer first. The result was a Bonus Built truck that put driver comfort foremost.

With its van-style "pie wagons," Ford was one of the earliest producers of light trucks. This beautifully restored 1912 "brass" Model T is owned by Dan Winters.

In the mid-fifties, when Ford restyled its trucks and created the immensely popular F-100 line, consideration for the buyer included adding a modern-design six-cylinder engine for greater economy. For the F-100 line's second restyle in 1957, Ford revolutionized pickup box design with the first all-metal, smooth-side box—the Styleside box. So popular was this wide box with its substantially expanded cargo space that Ford made it standard, and the traditional narrow box a special order item. Within a year or so, Chevy, Dodge, Studebaker, and International had all copied Ford's lead.

From the sixties on, Ford trucks have distinguished themselves in a number of mechanical and styling advances. In 1961, Ford introduced the first integral cab and box on a full-size pickup. Rugged Twin I-Beam suspension appeared in 1965, and the list goes on.

Ford has never been a copycat. From the

distinctive burbling sound of a flathead V-8's exhaust to the unique design of Twin I-Beam suspension, a Ford has always been a Ford. And that's a big part of what makes Ford trucks popular with collectors. "I'm a Ford person," you can say with pride. When the trucks we now restore were new, plenty of buyers felt the same way. Ford was a big seller—in many years, the top seller. And that's to our advantage because it makes Ford trucks still relatively plentiful. It's not scarcity or abundance by itself that determines value; it's desirability. Ford trucks were desirable when they were new, and they're even more popular with collectors today. All these characteristics add up to make Ford trucks enjoyable to own, relatively easy to buy and restore, and investments sure to rise in value. What more could you want?

The big part of a Ford pickup's appeal is that these vehicles have been somewhere in nearly every truck admirer's life. The first truck I drove was a Model A that had started life as a pickup and had descended to a farm runabout by the time it came my way. I used that dependable little A-model to haul lawn mowers for my high school lawn care business. When winter came and the pond near my parents' home froze sufficiently, my A and I would spend many pleasurable hours sliding and spinning on the ice.

Later, when I worked for Pan American Airlines in the Caribbean, I also spent time with Ford trucks. We used them as utility vehicles for whatever needed hauling, from people to pineapples. Some administrator whose job description must have read "Keep everybody's life dull and boring" had directed that the trucks be starved for fuel so that the top speed on the island's winding roads wouldn't exceed 30 miles per hour (mph). It didn't take much mechanical savvy, though, to realize that the crippling had been done by inserting a plate that had just a few small holes drilled in it, between the carburetor and intake manifold. Unbolting the carburetor and removing that plate returned the truck's performance to normal and made travel to and fro in that tropical paradise much more enjoyable.

Chances are you have memories of Ford trucks, too. This book's purpose is to help you turn those memories into a truck you can drive and enjoy.

The most pleasurable part of owning an older truck is driving it. I don't mean that driving an older vehicle is easier than piloting a modern car or truck; actually, navigating an older vehicle down the road is much more a full-time operation. It's the *experience* of driving an older truck that you're likely to find more pleasurable.

A vintage pickup has been my main local transportation since my sons got their driver's license with the result that more often than not I would find the only "car" left home to be the older pickup. But I figure that makes me the lucky one.

When I climb behind the steering wheel in that old pickup, it's as though I've entered a time machine. It's 1949 again—and in a way, it really is. When I stop the truck, people want to talk: "Nice truck you got there. Have to fix it up much? My uncle had a Ford just like that on his farm." And we chat. Usually those who come over to talk don't know me, but we don't feel like strangers. Time, that busy pace of things, seems to slow down and come nearly to a halt. Whatever errands those who stop to talk and I are on, seem forgotten. It seems to me that's more the way people used to live. In the memories I have of being a kid going places with my dad, there were those long intervals when he'd stop and talk—to passersby on the street, his friends, the salesmen at the new car dealership, the counter

Early-forties Ford trucks had a particularly handsome styling. That's me standing beside a one-and-a-half-ton stakebed owned by the Henry Ford Museum, Greenfield Village

The best part of owning an old truck is experiencing the friendships that form at shows and club events. This 1952 Ford stakebed has become a gathering spot for old-truck admirers attending the Somerset Gas and Steam Engine Pasture Party, an annual event in the Blue Mountain region of Virginia.

clerk at the drugstore. I don't see this happening anymore, except when I drive into town with my old truck.

A friend told me that he met an elderly acquaintance at the post office one day and asked, sort of over his shoulder, "How you been?" "Terrible," the elderly gentleman replied, then began to recount his nightmare of an illness that had kept him in the hospital and at home convalescing for the previous six months. This was his first day back in the wider world. "Good," my friend replied, thinking he'd heard the usual "Fine" and not having tuned in to a word of the man's grief.

We get like that, calloused through rubbing and being rubbed by people who don't have enough time for some real human contact. An old truck can put us in touch with those whose stories can make our life more interesting. If you're making a run to the store, why not travel in 1955 instead of the nineties. You'll find you've brought a little less frenzy into your own life, and into the lives of others as well.

So OK, you've decided to add a vintage Ford truck to your vehicle stable. If you are still searching for your collector truck, or are simply a Ford truck enthusiast, you'll want to read chapter 2 of this text for a quick review of the various models of Ford pickups starting with the Model T and proceeding to the sixties versions. There you will also learn the unusual features of the different models and what to look for to make sure your purchase is a sound investment.

If the truck you purchase is in its original used and abused state, chances are, you're asking yourself some questions about where and how to begin the fixing-up process. First, though, some even more basic questions must be answered. These include, What is your goal for your truck? What do you have available in the way of tools, equipment, and shop space? What financial resources are you willing to invest in the project? And perhaps most important, what amount of time can you commit to rebuilding an old truck?

Chapter 3 will take you through the process of answering these questions and helping you decide which path—restoration or repair—you will take in putting your vehicle back into respectable condition.

Chapter 4 discusses the tools and equipment you need to work on your project truck. Chapter 5 outlines the steps for disassembling and cleaning the vehicle. Chapters 6 through 16 cover the actual restoration or repair processes of various components and assemblies in Ford pickups. Finally, chapter 17 describes procedures for the proper care and preservation of the finished product.

Chapter 2

Ford Pickup Trucks Through the Years

1924–1927

The earliest light trucks were delivery vehicles, not pickups. Ford began building these deliveries—commonly called pie wagons because of the crescent shape cut away on the front of the body outlining the driver's compartment—shortly after introducing the Model T. It was not until the mid-twenties, however, that the pickup truck emerged.

Like those of other manufacturers, Ford's early light trucks—deliveries and pickups—were based on car chassis and used-car sheet metal. In fact, Model T pickups differed from a standard roadster only in that a small pickup box replaced the turtle deck "trunk." For those interested in owning a real antique, Ford Model T roadster pickups make fun trucks. The 20 horsepower (hp) engine with its splash oiling and spindly crankshaft, plus two-wheel brakes and wooden spoke wheels prevent these vehicles from traveling at highway speeds, but the legendary planetary transmission makes them unusual and amusing to drive. Even though these

Ford created its first pickups during the Model T series, by removing the back turtle deck from a roadster and bolting a compact pickup box in its place. The Model T pickup shown here is a 1927 vintage. All T pickups had open cabs.

early trucks are well over a half century old, plenty of Model T pickups can still be found—and prices have flattened out, which means you can buy a well-preserved Model T roadster pickup for about the price you would have paid for the same vehicle in 1960. The disadvantage here is that you're not likely to recoup more than that price should you sell the truck a decade or so from now.

For the last two years of Model T production—1926 and 1927—Ford updated the styling somewhat, offering a limited array of colors, optional steel spoke wheels, and a nickel-plated radiator shell. Except with the early brass Ts, for which a pickup wasn't offered, the end-of-production models are generally considered the most attractive, so if you have a hankering to own a T, a 1926 or 1927 model is recommended. Ford didn't make any major changes to the mechanicals throughout the T's production run, and parts, both original and new, are plentiful and relatively inexpensive. You will also find well-established clubs for T owners (see the Appendix).

Another advantage of owning one of these early Fords is that they qualify for low-cost antique license plates and insurance. The limited driving restrictions imposed by antique status are not a problem with the Model T, since these rather primitive car-pickups are "practical" only for parades and occasional short pleasure runs.

1928–1931

With the Model A, introduced in 1928, Henry Ford went upscale. Horsepower doubled to 40, and a manual three-speed transmission replaced the two-speed planetary unit. Other improvements included four-wheel mechanical brakes, an antitheft ignition lock, and "baby Lincoln" styling. For the first time, Ford offered both closed- and open-cab pickups. Light commercial models—the pickup and sedan delivery—continued to share passenger car chassis and running gear, as well as fenders and other sheet metal, but the cabs were distinct. The roadster pickup was also unusual in that the top didn't fold down; instead, it had to be removed for open-air

It is still possible to find restorable vintage Ford pickups in junkyards. This 1935 V-8 model, resting after years of faithful service, could be brought back to show condition by following the restoration sequences described in the chapters that follow.

When the Model A appeared in 1928, Ford had reworked the pickup so that it was available in either closed- or open-cab versions. Typical of Ford's policy of carrying over parts from one model to the next, doors for the 1928–29 Model A closed-cab pickup shown here interchange with those for the 1926–27 Model T coupes.

Sometimes you will find an older Ford pickup that's so well preserved it only needs some cosmetic touching up to be show-worthy. This 1939 model is right off the farm. Some bump work to the fenders, a repaint, and some trim work, and this truck would be a beauty.

A rare and appealing Ford truck is the Cab-Over-Engine model. The example shown here is 1938 vintage and owned by Bob Drake, who operates a Ford truck parts business.

driving. An aftermarket folding roadster pickup top is available from LeBaron Bonney Company.

Model A Fords are probably the most plentiful collector vehicles, and like the Model T's, their prices appear to have flattened. This means that you won't pay any more for a restored or restorable Model A closed-cab or roadster pickup than you would have a decade ago, but you probably won't receive a great deal more than you paid for the truck if you sell it a decade hence. So we're not looking at an investment here, but Model As are enjoyable vehicles to own and drive. Besides their simplicity and good looks, they have their own distinctive sound, unlike that of any other vehicle. They are also much more practical for longer cruises than are Model Ts. Equipped with an overdrive transmission, a Model A can cruise relatively effortlessly at highway speeds—just be aware of the limitations of the mechanical brakes!

Here, too, parts are plentiful. The two long-established Model A clubs have assembled excellent

Ford's 1940–41 car-inspired front end styling makes these pickups some of the best-looking prewar models.

mechanical data, including a book of restoration standards that is a thorough guide to rebuilding any Model A to factory-new condition. Owning and driving a Model A, particularly a roadster-style pickup, can be just plain fun.

1932–1939

In 1932, Henry Ford made history once again, this time with a low-cost V-8. Ford cars and pickups were given new styling—the most prominent feature of which was a painted radiator shell. The roadster pickup continued into 1933, but few examples of this model were built after 1929. Buyers apparently preferred the greater practicality of the closed cab.

As an economical alternative to the V-8, Ford offered a refined version of the Model A four-cylinder engine in its cars and light trucks through 1933. Engine upgrades included a fuel pump, a redesigned head and water pump, a counterbalanced crank on all 1933 engines but not all 1932 engines, and improved oiling. These improved engines can easily be fitted into a Model A and are desirable for that purpose.

Ford's pickup styling remained the same through 1934, with just a couple of minor, telltale changes, namely a new side hood insignia and chrome front bumper on the 1934 models.

For 1935, Ford restyled the pickup cab, giving it a more modern rounded and sweptback look. The attractive grille of the 1934 Ford cars now adorned the pickup truck.

Despite the generally dismal economic climate caused by the worldwide Depression, the thirties were a highly competitive era in the automobile industry, and to keep pace with the competition—namely, Chevrolet—Ford restyled its pickups again in 1937. New features included a split windshield, a cab with a higher crown, and front and rear bumpers as standard items for the first time. The appearance of these trucks, with their prominent oval-shaped grille with horizontal slats, continued more or less the same through 1939.

The 1938 and 1939 models, where the grille was more rounded across the top, have acquired the nickname "barrel nose" or "barrel grille." A new bed design also appeared on the 1938 models and was used for the next fourteen years.

1940–1941

With only minor changes to the cab—namely, elimination of the seam above the windshield—but completely new front fenders, hood, and grille styled to look like those of Ford cars of the period, the 1940 and 1941 Ford pickups became some of the most handsome trucks of all time. The headlights were now set into the front fenders, and the package was set off with a bright chrome bumper.

For 1942, and again in 1946 and 1947, Ford gave its pickups a more businesslike look. The grille on this series truck consisted of vertical teeth (missing from this example) that are often described as having a "waterfall" appearance.

A very unusual Ford pickup, also of the 1940–41 vintage, is this Australian-built Ute. Note that this is not a pickup in the usual sense, with a separate cab and box. Instead the side sheet metal for the cargo space extends from the abbreviated coupe body.

Here we see a Mercury pickup of 1942–47 vintage built by Ford of Canada. Although rather woebegone, this truck is identical in appearance to U.S. models, except for the horizontal grille bars and Mercury nameplates.

For the years 1937 through 1940, Ford pickups and other light commercials came standard with the 85hp 221-cubic-inch (ci) V-8 but could also be special ordered with the diminutive 60hp V-8. These tiny engines were underpowered for most work and would be a dog to drive on the highway today, but they are unusual—and their tiny appearance in the engine compartment makes them conversation material indeed. For 1940, the 60hp V-8 was improved, with hardened valve seats and a larger-diameter crankshaft, but the buying public showed that it wasn't interested in economy at the expense of performance, and Ford dropped the little V-8 that year.

In 1941, the economy option became a four-cylinder engine shared with Ford's new 9N tractor. Rated at 30hp, this tractor engine represented a real step backwards in truck performance, and very few examples were sold. Even rarer was the flathead six-cylinder engine also offered in Ford light trucks for 1941. Actually, the six was a very good engine, having the same horsepower as the V-8, and more torque. To fit the six into the relatively short engine compartment, Ford engineers had to redesign the front cross-member and add frame extensions on which to mount the radiator. The hood latch mechanism also had to be revised, as did the inner fender panels. A 1941 Ford pickup with the factory-installed six should be considered a very good investment.

1942–1947

Oddly, with war looming so close on the horizon, Ford extensively reworked its pickups again in 1942. A definite "truck" look replaced the design of the carlike front end. Added cargo space resulted from a 3in increase in the width of the bed.

Although the war soon cut off production, this truck returned, unchanged, in 1945 and continued to be built through 1947. The postwar versions eliminated the flashy chrome that had graced the front of the 1940 and 1941 models. Now the bumper, grille, and all trim were painted. Ford continued to offer the six as an alternative to the V-8, although most of the later trucks were V-8 powered.

Although more plentiful than the 1940–41 models, Ford's 1942–47 variant is far from abundant and makes a good play as well as show truck.

Ford produced a winner with its Bonus Built trucks, produced from 1948 to 1952. A wider grille opening with three prominent "teeth" marks the 1951–52 models. The example shown here is a 1952, recognizable by the single-bar hood spear.

1948–1952

The completely restyled Ford trucks that appeared in 1948 were called Bonus Built. Their primary bonus came in the form of what the ad copy called a Million Dollar Cab with Living Room Comfort.

Ad hype aside, owners of 1948–52 Bonus Built Ford trucks experienced many benefits over the offerings of previous models. An increase in cab height gave more headroom—handy for tall drivers and passengers or those who wore hats. An increase of 7in in cab width gave more room for that third rider. Moving the door hinges ahead 3in to give more space between the cowl pillar and the seat riser made for easier entry.

Other Bonus Built features included Level Action Cab Suspension (the use of rubber mounts to isolate the cab from frame flex), improved ventilation systems, and a one-piece windshield for increased visibility. Living Room Comfort referred to an adjustable seat that moved easily on roller bearings, and seat cushions constructed in a manner similar to that of an expensive mattress, for reduced driving fatigue.

By anyone's standard, the Bonus Built Fords were handsome trucks. The recessed grille and sweeping fender contours gave a streamlined frontal appearance. On 1948 models, the grille bars were chrome-plated. Silver (argent) paint replaced chrome in 1950.

Ford now used a combination letter-number system to identify its trucks by their load ratings. (International had introduced this identification scheme in the thirties.) All models used an *F* prefix: F-1 designated half-ton models; F-2 referred to medium-duty three-quarter-ton trucks; and F-3 meant a heavier-duty one-ton truck. The series continued through F-8 on the big trucks.

Only minor changes occurred between the 1948 and 1950 models. The F-series trucks received a number of alterations and improvements. Most noticeable was a restyling of the grille from the original five horizontal bars to a massive opening that extended to the edges of both front fenders and had a single horizontal bar straddled by three upright *dagmars* (cone-shaped protrusions). Additional changes could be seen inside the cab in the form of a restyled dash. Deluxe trucks now offered foam rubber seat padding and sound insulation in the headliner.

A big change came in the last year of the series, in the form of a completely redesigned overhead valve six-cylinder engine. The six's 101hp rating nearly equaled the 106hp rating of the flathead V-8.

Bonus Built Ford trucks are immensely popular with collectors for a variety of reasons. Their styling has a timeless, "classic" look. The V-8 engine is adaptable to a full range of hop-up equipment, all of which is readily available in reproduction form. Parts are plentiful, the simplicity of these trucks makes them relatively easy to restore, and the Living Room Comfort cab gives a reasonable standard of driving comfort.

1953–1956

Ford Motor Company celebrated its fiftieth anniversary in 1953. The car line-up was a year old, so the new truck and farm equipment lines held the spotlight. These Golden Anniversary pickups represented a clean break from their Bonus Built predecessors in every department except engine choices. Along with a totally restyled cab with its smartly slanted windshield and short roof cap, Ford's F-100 half-ton pickups sported a longer (6 1/2ft) and taller (20in) box that would be used into the eighties. For the first time, the pickup tailgate dispensed with

Bonus Built Ford pickups look good from any angle.

Ford built another series of winners with its popular 1953–56 F-100 series. With a V-8 engine and overdrive, these trucks could cruise down the highway as effortlessly as a car.

the Ford script, instead displaying the Ford name in block letters.

Probably the most noticeable difference between the F-100–and F-1–series trucks was the increase in window area. In windshield area alone, driver visibility increased 55 percent. Not only did a wider rear window and taller side windows give near-panoramic vision, but a lowered cab beltline made possible the trucker's favorite pastime: resting an elbow on the window ledge.

Another significant design change, but one that takes a sharp eye to spot, is the shorter cab–to–front axle distance. This feature, which Dodge introduced on its B-series trucks and International adopted on its L-series, gave the new Ford pickups a shorter turning radius for better maneuverability.

Although the flathead V-8 carried over into 1953—along with the new overhead valve six—starting in 1954, a downsized version of the reliable Y-block V-8 that had powered Lincoln cars and Ford heavy-duty trucks since 1952 topped the engine line-up. Other changes during this four-year styling series included annual grille modifications; the advent of tubeless tires—adopted industrywide in 1955; and the introduction of the wraparound windshield on the 1956 models.

In a move that is somewhat puzzling to collectors, the new series model designators ran F-100, F-250, F-350, whereas the preceding series identified its models as F-1, F-2, F-3, and so forth. Why not F-100, F-200, F-300? The reason is quite simple: Above the light-duty half-ton pickup, Ford created a new heavier-duty truck that had a load-carrying capacity between that of the former three-quarter-ton and one-ton models. This truck became the F-250. Likewise the F-350 fitted between the former F-3 and F-4.

These F-100 series trucks built between 1953

Learning About Your Truck

Before you set out to restore your Ford truck, it helps to know as much about that vehicle as possible. Identifying the year and model is usually pretty easy. Unless the title has been messed up over the years, it will have the truck's year. The model series should be recognizable from the photos and descriptions in this chapter. But suppose you want to verify points of authenticity—such as whether or not your truck has the correct engine, whether the wheels are the original size, what the color options were, whether a chrome-plated grille was an option, and other details of this sort. How do you go about getting this information? Basically, three good sources are available. One is books, another is original Ford sales and service literature, and the third is club publications.

Much of the authenticity information for restoring the 1950 Ford F-1 pickup seen in many of the photos in this book came from Paul McLaughlin's *Ford Pickup Trucks: Development History and Restoration Guide, 1948–56*. For collectors and restorers of later-model Ford pickups, McLaughlin has written a companion book, *Ford Pickups, 1957–67: How to Identify, Select, and Restore Ford Collector Light Trucks, Panels, and Rancheros*.

Owners of earlier Ford pickups will find the photo book *Ford Pickups, 1932–1952* by Mack Hils a helpful guide to restoring their vehicle. Although this book has very little text, most of the photos are from the Ford archives and show trucks in their factory-new condition. Captions with the photos point out many details that go into making a show-winning restoration.

For production figures and specifications, the *Standard Catalog of American Light Duty Trucks, 1896–1986* by Krause Publications—publisher of *Old Cars Weekly*—is a good, comprehensive source book.

Sometimes we're looking for more than data; we want the story behind the truck—like how Ford's marketing research led to the creation of the Bronco. That tale and a wealth of description and historical detail are provided in the *Heavyweight Book of American Light Duty Trucks, 1939–1966*.

If you want to see what your truck looked like when it left the factory, check specifications, and perhaps learn of options available, and to get a sense of the color selection, dealer sales literature is the answer. These color pieces are available from literature venders whose wares you will find at major car shows and swap meets. You can also learn important details about your truck from factory service manuals.

One of the best ways to learn about your vehicle, and get in touch with other Ford truck owners and restorers, is by joining a club serving owners of vintage Ford trucks. A club for owners of all years of Ford trucks is the Light Commercial Vehicle Association. Besides the historical features and technical data that appear in its newsletter, this club has technical advisers who can help you with parts information and questions you may have about your truck.

If your truck is a Bonus Built model, the *'49-50-51 Ford Owners Newsletter* can be a helpful information source. This publication, which operates much like a club, frequently prints Ford truck service letters and features a technical column for Ford trucks. You will find addresses for both the Light Commercial Vehicle Association and the *'49-50-51 Ford Owners Newsletter* in the Appendix.

and 1956 have long been popular with collectors. Street rodders, especially, have a fondness for them. As a result, working F-100s in restorable condition are somewhat hard to find. But if you're lucky enough to locate one, these early pickups are quite easy to renew. Body and mechanical parts are plentiful in reproduction form. Because of the street rod interest, an abundance of conversion equipment is also available to bring these trucks up to modern handling and comfort standards.

1957–1960

When Ford stylists set about designing the 1957 truck line, they appear to have started with a clean sheet of paper. Slab sides replaced the "fat-fendered" look. Every line was either straight or angular.

Trucks of this design, which were built from 1957 through 1960, are just beginning to catch on with collectors. The rectangular look lacks the romance of the earlier models' classic styling, but the "slab-sided" models have their own collection of pluses. These were the first trucks to offer a true wide box—standard equipment, no less. The old-style narrow box now had to be special ordered. Running boards disappeared; a first for Ford. The engines were more powerful, the interiors livelier. Following Ford's pioneering safety steps, pickups also featured deep-dish steering wheels, double-grip door locks, a padded dash on the Custom models, and optional dealer-installed seatbelts.

The major drawback to this series is the difficulty of repairing rust to the Styleside box. Reproduction sheet metal is not available, making serious rust repair troublesome.

1961–1966

Ford's next major truck restyle came in 1961. For all of their successes, these new F-series Ford pickups had one design quirk that didn't make it. Some call it a flaw; to others, it's a collector's prize. The item in question is the integral cab-box Styleside pickup.

Ever since Ford had first "invented" the pickup truck by bolting a cargo box on the rear of the chassis behind a roadster tub, truck builders had more or less assumed that pickups had to be of the two-box design—the forward box carrying the driver and passengers, and the rear box for the cargo. Sure,

Like the series it replaced, the F-100 looks good from all angles. The example shown here is a 1953 model.

earlier designs had fused the two boxes into one—the Ranchero being a prime example—but not on a full-sized pickup. Ford departed from the two-box tradition, and from 1961 through 1963, its Styleside trucks consisted of just one box, with the pickup bed and rear fenders being an extension of the cab.

The integral cab-box pickup was a great idea. By eliminating the space between the cab and bed—this was done by making the rear of the cab the front of the bed—a longer cargo area could be achieved without lengthening the wheelbase. Also, the design looked great. And because they didn't have two boxes to twist and turn independently of each other as they rolled over bumps, the one-box trucks gave a smoother, quieter ride. But in time, problems emerged. Rust brought many of these trucks to an early end. There were also reports that when the trucks were loaded, the doors would bind and couldn't be opened. Talk like this began scaring buyers away.

Ford didn't box itself into a corner, however, and from the start of this integral design series, the traditional narrow-box Flareside pickups were offered. Then in 1963—the third year of the Styleside series—Ford quietly offered two wide-box models: the integral design, and a separate cab-box pickup that used the old Styleside box from the slab-sided series. The old box looked like what it was—an afterthought—but it addressed buyers' fears about the integral design model.

In 1964, Ford quietly eliminated the integral cab-box construction and reverted to the two-box design. The new Styleside pickups had the same load length as before, and the wide box now became available on Ford's 4x4 models. Ford continued this F-series styling through 1966, with the major improvement in 1965 of Twin I-Beam

Ford of Canada marketed its M-100 pickups under the Mercury nameplate. In all other ways, these trucks were identical to their U.S. F-100 counterparts.

For 1958, Ford introduced a wide, smooth-side box—an industry first. The narrow, fendered box was now an option. The example shown here is a 1960 model owned by Debbie Schissler.

Ford Bonus Built Serial Numbers

The serial numbers for the 1948 Ford Bonus Built pickups start at 88RC-101 and run to 88RC-139262. The first 8 is the model year (1948), the next 8 and the R mean the eight-cylinder engine, and the C means the half-ton class. Of the three years for the Bonus Built series, 1948 was the easiest to identify just by virtue of the serial number. The 1949 trucks started at 98RC-7388 and up, and the 1950 models continued the 1949 serial numbers.

Besides the serial numbers, some subtle differences between Ford's 1948, 1949, and 1950 F-1 models may help identify a Bonus Built truck's year. For 1948, the wheels were painted the body color, whereas in 1949, they were painted black; the vent window bar was chrome in 1948, black in 1949; and the painted grille bars had a small red pinstripe in 1948, which was deleted in 1949. The F series emblem on the cowl sides can help distinguish a 1949 from a 1950 model. If these emblems are die cast and chrome-plated, the truck is a 1949 model. If they are stamped stainless steel, it is either a very late 1949 or a 1950. For 1950, the main difference was that the gearshift lever on the three-speed was moved to the steering column.

Ford introduced its compact Econoline pickup in 1961. Powered by an economical six-cylinder engine, the Econoline made an ideal light delivery vehicle.

Ford's full-sized pickups for 1961 also received new styling. Prominent features of this series, which was produced from 1961 to 1966, were a less angular hood and body edges and stylish wheel cutouts. A 1963 example is pictured here.

For 1967, Ford pickups underwent another styling change, which returned to the more angular look. The example shown here is a 1967 Mercury owned by Victor Buck.

independent front suspension (IFS). The advantage of IFS was a greatly improved ride, which, combined with ever-more-powerful six-cylinder and V-8 engines, put Ford trucks solidly in the modern era.

Econoline pickups and vans were introduced in this period, in 1961, as was the Bronco, in 1966. Ford F-series pickups of the 1961–66 styling series that are gaining the strongest collector interest are the 1965–66 models with Twin I-Beam front suspension and the 1961–63 integral cab-box design. Parts, including sheet metal, appear to be more plentiful for these models than for their slab-sided predecessors, and the more rounded design seems to have greater appeal.

1967–1979

In 1967, Ford trucks underwent another styling change that resulted in more angular lines. Gone were the vestigial fender outlines around the wheel openings. With modifications to the hood and grille, Ford trucks would keep the same cab and box design through 1980.

The major development during this styling series was the introduction in 1975 of the F-150. This heavy-duty half-ton with its 6,050lb gross vehicle weight (GVW) rating enabled Ford to bypass emission standards. Until 1979, F-150 models continued to be exempt from mandates requiring unleaded fuel and were not fitted with catalytic converters.

Because these were strong working trucks, few have been retired to collector status. If you own or decide to purchase a well-maintained, solid truck from this series, however, you may want to set it aside for collector use. Chapter 10 discusses a 1969 Ford camper special that is undergoing its second repaint in preparation for an easier life as a truck that sees fewer miles hauling a camper and spends more time touring to old-car and -truck shows.

Ford successfully stretched its 1981 restyling into the nineties, winning a series of hard-fought sales battles with Chevrolet. Bill Randall, owner of the truck shown here, operates the Horn Shop, a restoration shop for vintage car and truck horns.

1980–1991

Ford trucks in the 1980–91 styling period are still too new to be looked at seriously by collectors. But if you own one of these trucks, and would like to keep it for many years to come, then you'll find plenty of tips and techniques in the following pages that will help you recondition and preserve your vehicle.

Chances are you've got some good memories of Ford trucks, new or used models you've come across in life.

Chapter 3

Rebuilding or Restoring

The repairs and cosmetic upgrading needed by most older pickups can be approached in two quite different ways. One method is to make whatever repairs are needed—rebuild the engine, brakes, front end, and steering; do the bodywork and refinishing; replace the flooring in the bed; and redo the interior—but never take the truck entirely apart. This approach can be called rebuilding. Its goal is to produce a good-looking, drivable truck at the minimum expense and time.

The other method is to take the truck apart—right down to the last nut or bolt—and then rebuild everything, strip the finish to bare metal, sandblast or chemically remove all surface rust, and repair any rusted-out metal. This is called *restoration*. The goal of restoration is to bring your truck as close as possible to what it was like when it rolled off the assembly line, but with the quality finish and artisanship that could be expected on a Rolls-Royce or other skillfully built vehicle.

Deciding whether to Rebuild or Restore

The decision of whether to rebuild or restore should be based on thoughtful consideration of the pluses and minuses to each approach. To help you make that decision, the merits of both approaches will be examined in five categories: (1) your goal for your truck; (2) your truck's intended use; (3) the amount of money you are able—and willing—to put into the vehicle; (4) the amount of time you have to spend on the rebuilding or restoration process; and (5) your patience or endurance. After you have weighed both approaches by these criteria, you will know which direction to take.

Rebuilding

Although your truck will be tied up for short periods during the rebuilding process—while the engine is being overhauled, for example—the piecemeal rebuilding approach allows you to drive the truck during most of the time that you are upgrading its appearance and reliability. Another benefit of rebuilding is being able to spread the cost of most upgrades over as long a period of time as is necessary, while still using and enjoying the truck. Rebuilding also takes less shop space than restoration because the truck is never taken completely apart. If this is your first experience working on a vintage vehicle, you're not as likely to lose interest in the project if you take one rebuilding step at a time, as you would be if you took the truck completely apart and scattered all of its pieces over your garage, and then had to figure out how they all went back together.

If you're fortunate enough to find—or own—an older pickup that is still in good original condition, you'll be much further ahead to do just the necessary cosmetic and mechanical repairs than to take it all apart and restore it like new. Any truck can be fixed up with new parts, but an older vehicle that has

If you're lucky enough to find a vintage Ford truck as well preserved as this 1947 Cab-Over-Engine model owned by Bob and Jean Peck, only minor repair and refinishing is needed. This truck's rebuilding consisted of replacing the driver's seat covering, performing minor engine work, and doing some touchup detailing.

been preserved in its original condition has a special charm.

A goal of rebuilding, especially if you are starting with a truck that's in fairly good condition, is to bring your vehicle back to what it looked like when it was two or three years old. It won't be so perfectly manicured that you'll have to shoo the neighbor kids away, and you won't be afraid to leave it in a parking lot. People will still stop and tell you how good your truck looks, and many will probably tell you that your truck reminds them of one they owned. That's what rebuilding accomplishes. It doesn't build a monument; instead it gives you a truck you can enjoy. For many collectors, that's the kind of old truck they want.

The steps to rebuilding a "typical" older pickup include removing the fenders and pounding out the dents; repairing rust—typically found in the cab cowl and rear corners; usually overhauling the engine; and almost always reworking the brakes, replacing the wiring, upgrading the interior, and repainting the exterior. Numerous other jobs may also be called for—from replanking the box to replacing the window glass.

The one major drawback to the rebuilding approach is the likelihood that you will later find something you wish you had done more thoroughly. For example, if you only repair visible body damage, you're likely to overlook rusted inner body panels and spongy cab supports. Repairing this hidden damage is just as important in rebuilding a truck as it is with restoration, but not all problems are likely to get noticed during the appropriate phase of the work. Later, when you're hooking up an exhaust system, you may glance up at the cab supports and suddenly realize that the cab is held up by Swiss cheese. Replacing the support brackets is a lot harder when nearby surfaces have just been freshly repainted.

Restoring
Unless you ship your truck off to a professional restorer, you can plan on a ground-up restoration taking from one to three years—possibly even longer. Gary Edgerly, an auto body instructor at a technical college who does restorations in his home shop on a part-time basis, took more than a year to complete the Ford F-1 pickup seen in many of this book's photos. Edgerly has the advantage of years of training and experience, and an extensive tool stock. The same work could take the first-time amateur restorer considerably longer.

During restoration, the truck will be spread all over your shop or garage. Some parts may even sneak their way into your living area. I have a speedometer for my truck sleeping in a desk drawer; a friend went so far as to store the headlights from his thirties-vintage truck behind the living room sofa—he was afraid they'd get banged up in the garage. In these circumstances, there's always the danger that you'll misplace some of the parts; you'll need to develop a workable inventory system. I couldn't find the throttle linkage to my truck until after I brought home a replacement from a scrap yard. Of course, you always run the risk of getting

---

Preparation Steps

1. Decide whether you intend to follow the rebuilding or restoration approach.
2. If you are repairing your truck, list the work needing to be done.
   a. Overhaul the brakes.
   b. Replace the wiring.
   c. Rebuild the engine.
   d. Repair sheet metal damage.
   e. Repair the finish, etc.
3. If you are restoring your truck, document it with notes and photos.
   a. Record the weather stripping location.
   b. Write down the original colors.
   c. Indicate the pinstripe location, if any.
   d. Identify original and nonoriginal accessories.
   e. If you are repairing your truck, list the work needing to be done.
4. If you are restoring your truck, begin researching it.
   a. Look up the coding on the truck's data plate.
   b. Determine the truck's original color.
   c. Learn the correct coating for bolts and fasteners.
   d. Find out if the engine is correct and original.
   e. Determine if the drivetrain is original.
   f. Check whether accessories are original; learn what accessories were available for the truck.
5. Determine standards.
   a. Will the truck be modified or stock?
   b. If the truck will be modified, in what ways: brakes, electrical, suspension, comfort?
6. Estimate costs.
   a. Take a tool inventory; establish a budget for needed tools.
   b. Estimate the cost of professional services.
   c. Factor in travel expenses.
7. Purchase parts and service manuals, sales brochures, reference materials, and resource books.
8. Develop a supply network.
   a. Order parts supplier catalogs.
   b. Ask the counter clerk at your local NAPA store about parts availability for your truck.
   c. Find out scrap yard locations.
9. Decide where to begin.
   a. Rebuilding makes vehicle safety the first concern.
   b. Restoration starts with complete disassembly.

discouraged by all those piles of parts.

The benefit of restoring is that every part of the truck that needs attention gets it. When the restoration is finished, the truck should be better-than-new. At that point, you will face the restorer's dilemma: Can you drive and enjoy your truck—and endure the inevitable scratches, chipped paint, oil film on the immaculate engine compartment, and grime on the glistening undercarriage—or must the truck now be an object to be looked at and not touched or driven?

Before you begin a frame-up restoration, you should establish a plan that includes your projected and fallback time schedules and budgets. I advise that your schedule include an alternative timetable to keep discouragement in check during those inevitable setbacks and obstacles. Virtually everyone who works on an old truck experiences unexpected problems—some amusing, some downright devilish.

One troublesome setback in my rebuilding experiences was an improperly ground crankshaft that caused the rods to bind so tightly that the engine wouldn't turn when reassembled. To free the engine, the unit had to be disassembled again and I had to find an engine artisan who fitted the rod bearings to the crank—not normally an advisable procedure, but one that old-time engine specialists are able to do successfully. The alternative would have been to have the crank reground—decreasing the bearing size still more—and then buy another set of larger oversize bearings.

A friend told of placing the just-refinished tailgate to his pickup in the sun in front of his shop to dry. Kids playing baseball in a nearby field hit a fly ball that landed—you guessed it—smack in the middle of the tailgate. In addition to repainting it, this restorer had a baseball-size dent to work out of the tailgate that a few moments before he'd thought he'd brought back to perfection.

When setbacks happen, instead of getting down on yourself and crying, "Why me?" it's a lot healthier and less stressful to say to yourself, "These things come with the territory," and move on.

For setbacks and obstacles like parts that break in the process of fixing them or work that wasn't expected, your restoration budget should include a contingency; I'd advise 30 percent. If the extra isn't needed, you'll be money ahead when you're finished. But if you don't allow for unanticipated extras, you may fall short of resources to finish your truck—just when the end is in sight.

Most first-time restorers begin by taking their

It's probably about this stage that you'll start seriously thinking about whether your truck needs the full restoration approach or if you will take the step-by-step rebuilding approach.

vehicle apart. This is the wrong approach. It's true that the vehicle will have to be disassembled, but this is *not* the first step. You should begin by researching your truck as thoroughly as possible. The research step is discussed in more detail later in this chapter because it is also the starting point if you decide to take the rebuilding approach.

Part of what's involved in the research—which means learning all you can about your truck—is documenting your truck as completely as possible. You begin by gathering as much information as you can find on the vehicle itself. Note the codes on the data plate; the original color—usually visible underneath the hood, on the back of the cab behind the seat, and in other places that aren't likely to be repainted; engine and chassis numbers; and details such as mileage, accessories, tire sizes, and the like.

Documenting is also done by taking clear color photos of the truck from every angle. A quality camera, preferably a 35 millimeter (mm), should be used for the photos. If you're not a shutterbug and don't have a 35mm camera with a telephoto lens for close-up shots, ask a friend whose hobby is photography to take the photos for you. Offering to buy the film and treating the photographer to a dinner should be a good trade for a half-hour or so photo session.

Be sure the photos are taken from every angle, front and back, top and bottom. Note such details as the placement of the weather stripping around the doors and the width of striping on the wheels or body reveals, the presence or absence of welting between the fenders and body—everything you can think of. From your research, you will also make a list of everything about the truck that you find to be unoriginal. Examples include incorrect taillights—from a hardware or auto supply store; signal lights on trucks not originally so fitted; a replacement engine; an incorrect box; jerry-rigged wiring; missing hubcaps; and so on.

Before taking the truck apart, make a list of all the items you know you'll need to replace—because they're missing, worn-out, damaged beyond repair, or incorrect for your truck. Missing items may include the floor mat or the correct taillight. Chances are, you'll find plenty of parts that are worn-out. One likely example is the exhaust system. The radiator could well be corroded beyond repair. One or more

With rebuilding, the box is usually removed but the cab is left on the chassis.

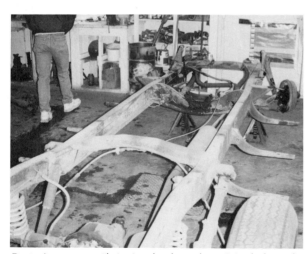
Restoring means "bringing back to the original, from the frame up." You will remove all body parts, exposing the frame, then remove and disassemble the suspension and running gear so that the beginning point—as at the factory when the truck was built—is the bare frame.

As can be seen in this view, a frame-up restoration is needed to do a thorough cleaning and renovation of all chassis parts as well as to gain access to hidden areas of body sheet metal.

of the fenders, or the box, may fit this category as well. An incorrect item could include the engine. It's not unusual for a truck that has seen plenty of work to be running on its second or third engine.

If you are more concerned about comfort than about authenticity, you may also want to make an upgrade category, in which you can list items such as the parts necessary to convert to a 12-volt electrical system, a modern radio, and an air conditioning unit. Make the list as complete as you can without taking the truck apart. Then you can begin to lay out the restoration timetable and draw up a budget estimate.

Determining Standards

If you plan to enter your truck in show competition, make sure every detail, even the location of the door weather stripping seam, matches the way it was when the truck rolled off the assembly line. Restoring an older truck to original standards requires extensive research. If your truck has been repainted—and chances are, it has—you'll want to determine its original color. If you would prefer another color, you'll need to find out whether it was offered for your year and model truck, and if not, what the other color options were. You will also need to research original engine color schemes. If pinstripes were applied to the reveal moldings around the cab, you'll need to learn their location, size, and color options. Sometimes the location of the stripes will show up on the bare metal after the paint has been stripped or sandblasted off. If not, you will need to determine where the stripes were placed by looking at a truck that still has its original finish or at a correctly striped show truck.

When restoring a truck for show competition, you will come to appreciate such seemingly irrelevant details as the proper style cap on the tire air valves, the correct spark plugs with painted or plated bases, the original style markings on the door and windshield glass, the right coating for the bed wood. Not paying attention to details of this sort may cost you the trophy you've worked so hard to win.

Show competition isn't the only reason for restoring a truck to original standards. Preserving originality can also be personally satisfying and is the best insurance for maximizing your truck's value. Anyone can rework an older truck with new parts and coat it with any color her or his heart desires, but it takes stamina to track down original parts and to stick consistently to the original colors and overall appearance.

If you are rebuilding your truck so that it can be driven and enjoyed on the highway, original standards are not so important and you may even be willing to make modifications that will increase your driving comfort and bring the truck up to modern highway standards.

A complaint of most owners of older trucks is the low-speed rear axles installed in the days when a pickup may have seen the open highway only once or twice a week, taking its owner to town. A thirties- or forties-vintage pickup is likely to feel uncomfortable at speeds much over 50mph. One way around this may be to install a car rear axle. Another is to add an overdrive transmission. The torque tube driveline Ford used in its half-ton trucks through 1942 didn't allow overdrive to be offered as a factory accessory, but installing this feature is a regular occurrence at Overdrives, a business specializing in retrofitting an overdrive transmission to any vehicle, including light trucks. The benefits of an overdrive transmission are discussed in chapter 8.

Converting to a 12-volt electrical system is another popular upgrade. If you make this change, you'll have plenty of capacity for power-hungry accessories such as air conditioning and a modern radio-sound system. The higher voltage will also spin the starter faster, making the engine start more easily. This upgrade can be done so subtly that the 12-volt battery will be the only clue that the truck isn't stock.

Estimating Costs

Whether you decide to restore or rebuild your truck, you should prepare an estimate of the expenses you expect to incur. Include amounts for parts, labor, transportation, and supplies. You can estimate parts costs by going through a catalog from a major Ford pickup parts vender like Dennis Carpenter Ford Reproductions (see the Appendix for the address) and listing all the items you think may be needed for your truck, pricing out each item, and adding up the total amount. In actuality, you will probably shop around for parts, finding some at

Among the information sources needed to restore a vintage Ford pickup are shop manuals and catalogs from parts suppliers.

Disassembling a truck for restoration typically begins with removing the front clip (the hood, fenders, and grille assembly). This exposes the engine, which can be removed next. Chrystal Edgerly

With the front stripped to the frame, disassembly can move to the back of the truck. Here the running boards and rear fenders have been removed. Chrystal Edgerly

swap meets, others at auto supply stores, and still others at salvage yards, rather than place an entire "shopping order" from a single vender's catalog. On a large order, however, it's sometimes possible to negotiate a discount. It's unlikely that your parts bill will be less than the estimate, because of needs you can't yet anticipate. The actual bill will almost certainly exceed your estimate—the question is, by how much. I suggest a 20 percent "fudge factor."

The charges you will pay others to do your truck's bodywork, mechanical work, or trim work can grow very large, very fast. For this reason, most collectors try to do as much work on their vehicle as they can. The problem many first-time rebuilders and restorers encounter in trying to redo their trucks themselves is the lack of proper tools and the skills to use them. Bodywork is an example. You can't do much in the way of metal repair without a welder, and even if you invest a few hundred dollars in one with a suitable capacity, the tool isn't much use until you know how to operate it.

In estimating labor costs, make a list of the jobs you want to do yourself; then beside this list, indicate whether your current tool assortment will be adequate for this work, or if additional tools and

The bed comes off next. Note that this truck has been fitted with auxiliary coil helper springs—a sign, perhaps, that the rear leaf springs have weakened and will need to be re-arced or replaced. Chrystal Edgerly

skills are needed. For example, most of the mechanical repair on a vintage Ford pickup can be done with a 1/2in drive socket set, assorted wrenches, pliers, screwdrivers, a file, a hammer, a chisel, a hacksaw, a hydraulic jack, jack stands, and a few specialty tools like a gear puller. But for an engine overhaul or bodywork, you will need a larger stock of specialized tools and equipment.

Both the basic tool set and desirable specialized tools are discussed in chapter 4; what's important to consider here is how large a tool investment you are willing to make to do as much of the work on your truck as possible. If you find working with your hands to be a therapeutic pastime—and many do—then investing in tools is not only a great savings, but healthful as well. And there's an added bonus: Once you've made the initial investment, the tools are yours to use—probably for a lifetime. If you plan to do as much of the work on your truck as possible, put tools and training costs in the labor expenses category.

You will always need to hire out some work, and for this, you should get estimates. In many cases, shops will be reluctant to quote a price for mechanical or repair work on an older vehicle, not wanting to be held to the quote when unforeseen problems develop. You can explain that for now, you're just looking for a ballpark figure, which will probably be on the high side. In the labor cost category, you will include specialty work such as chrome plating and jobs better left to experts, like instrument gauge repair and windshield replacement.

Labor costs can be shaved in other ways, besides doing the work yourself. One alternative is to share work with friends. The times I've been able to do this have been among my most enjoyable restoration experiences. Once a friend and I both needed to rebuild the steering and front end on our Model A Fords. So we took apart the front end on each of our vehicles; determined the parts we needed; pooled our orders and got a small discount for placing the one larger order; and then set up an assembly line operation for cleaning, refinishing, and rebuilding both front end and steering assemblies. This approach was much more efficient than if we had overhauled each vehicle separately, but more important, we learned from each other and had a great time in the process.

Another cost-saving alternative, sometimes

available, is to have major mechanical work or bodywork done by students in a vocational-technical (vo-tech) program. Admittedly some risks are involved here. Parts can be lost, and the quality of the instruction will determine the outcome of the finished product. I teach at a technical college that trains students in engine rebuilding, bodywork, machining, and similar trades, and have consigned my trucks to these programs for work and painting, plus some engine and other mechanical work. The process is not very speedy, and problems do arise, but any lower-quality work—such as a heavy orange peel texture that appeared in the paint near the bottom of the driver's door on one of the trucks—has always been made right. The big plus to having a vo-tech program assist in a vehicle's rebuild or restoration is the nominal labor charges. A hoped-for offshoot is giving the students an appreciation of older vehicles.

Your expense estimate also needs to include a listing of supplies. This category covers everything from sandpaper and painting materials to masking tape; miscellaneous nuts, bolts, and fasteners; weather stripping glue; plastic sheathing to protect parts from dust; sand for sandblasting; reference literature, including sales brochures and service and parts manuals; in short, all the odds and ends you will need to complete your truck's restoration or rebuilding. You really can't accurately project supply expenses in advance. You should allow $400 for paint and at least half that amount for the manuals and other books you will use in rebuilding or restoring your truck. Figure in another $400 to $600 as a ballpark figure for the supplies you are not able to itemize.

The last category in your budget estimate is for transportation expenses. These may include hauling the cab to a chemical stripping facility; traveling to scrap yards and swap meets to hunt for parts; contracting with United Parcel Service (UPS) or a freight company to ship parts you may order from venders; and going to look at original, restored, or "in-process" trucks. You may think of these travel expenses as part of the enjoyment of your hobby, but they are also part of the cost of restoring or rebuilding your truck, so you might as well project some figures and tally them in. Whether or not you will be using your truck in any business capacity, be sure to keep a log of travel and related expenses to use in calculating an accurate appreciation figure

To lighten the cab in preparation for its removal from the frame, the doors are taken off and the interior is gutted.

Although it has a lot of surface rust, the cab on this Texas truck is in solid condition. Chrystal Edgerly

should you decide to sell the vehicle sometime in the future.

When you add up all the expense estimates, you may want to take a deep breath before you hit the total key on your calculator. Even rebuilding an older truck can be a more costly undertaking than you would think—and restoring is likely to run considerably more. But don't let the final figure sour you on sprucing up your old truck. You'll find that some of the costs can be trimmed, and the expenses will be spread over the duration of the project. If restoration seems outside your budget, you might consider the rebuilding approach.

Gathering Information

Certainly by now you're itching to attack your truck, wrench in hand, and start taking things apart. Just a bit more patience is needed. Before you scatter your truck all over the garage, you should acquire a set of shop manuals; a Ford parts manual for your year and model truck is also highly desirable. These texts are available from the literature dealers listed in the Appendix, and also can often be found at swap meets. While you're at it, order catalogs from several parts venders specializing in vintage Ford trucks. A suppliers listing, with addresses, can also be found in the Appendix.

You will use the shop manual to take the truck apart as well as put it together. By following the disassembly steps in the manual, you'll find that things dismantle easier—and you're less likely to break hard-to-replace items. Although shop manuals will provide quite detailed disassembly and rebuilding instructions, working on a vehicle that is twenty to fifty years old is different than overhauling the same vehicle when new. The rebuilding instructions in this book will take those differences into account, and will guide you through difficulties you may experience and provide tips for making old parts work like new.

It's also advisable to have a copy of the *Motor* or *Chilton's* manual covering your year and model truck. To get an understanding of how something operates, and to get a full understanding of the disassembly and reassembly sequence, it is often helpful to be able to read the instructions from more than one vantage point. The parts manual and parts

Four people can pick up and carry away the cab. Chrystal Edgerly

---

Rebuilding versus Restoring

	Rebuilding	**Restoration**
Goal	Make a reliable, running truck that looks as it might have when originally in use.	Bring the truck to showroom or better-than-new condition.
Approach	Start with a well-preserved truck.	Start with a truck in any condition.
Advantages and disadvantages	Costs are spread over as long a period of time as necessary.	A lack of funds at any critical stage will delay completion of the project.
	The truck can be driven most of the time while repairs are in progress.	The truck will be inoperative through most of the restoration period.
	Less shop space is required than with total disassembly.	A space at least the size of a 2-car garage is required.
	The hobbyist is less likely to lose interest than he or she would be with a restoration.	For the average hobbyist, a frame-up restoration will take 1–3 years.
Results	The rebuilder will likely find some things she or he overlooked or wishes had been done more thoroughly.	Every part needing attention will get it.

venders' catalogs will help you identify any parts needing replacement and will help you locate sources for these parts.

To get a sense of what your truck looked like when new, scout out and purchase sales literature for your year truck. These brochures will show original engineering and styling features, upholstery style, engines, and colors. The sales brochures will help you spot changes that have been made to your truck over the years and will guide you toward its authentic preservation. You should also purchase one or more of the Ford pickup reference books listed in the Appendix.

Shopping for Parts

The more quickly you develop reliable parts sources and tap into a good parts network, the smoother your parts hunting will be and the less likely you will be disappointed by inferior quality. The supplier listing in the Appendix was compiled from recommendations by many Ford truck restorers. But in this list, it pays to shop around. A good policy is to order catalogs from several suppliers, then compare prices, and place small orders at first to test the service. When you are satisfied that you've found a supplier whose service and quality you can trust, then you have a green light to go ahead with larger orders.

In shopping for parts, don't overlook nearby auto parts stores. NAPA (National Automotive Parts Association) auto parts stores, a nationwide network of independently owned stores linked to a massive warehousing system, have probably the most complete stock of parts for older vehicles. Not only are parts purchased from a NAPA outlet usually less expensive than similar items from a specialty supplier, but you will be charged no separate handling and shipping fees. In most cases, the local NAPA store can have a part within 24 hours after you place your order, if the item is not already on the store's shelves. If your area doesn't have a NAPA store, check parts availability at other auto parts outlets.

Whether you're rebuilding or restoring, you'll

Now the mechanical assemblies, such as the rear end and transmission, can be removed, cleaned, and overhauled. Chrystal Edgerly

also want to explore that most enjoyable parts source, the auto salvage yard. Unfortunately old-time "junk" yards filled with vintage vehicles are fast disappearing. Lists with locations of potential parts "gold mines" appear in *Cars & Parts Annuals*, *Hemmings Vintage Auto Almanac*, and occasionally club publications.

Ford trucks were such big sellers that most parts are quite easy to find. Another plus for the Ford truck collector is the interchangeability of many mechanical parts between Ford cars and pickups. To find out what parts can be interchanged, you will need to consult the *Hollander* manuals. These books are the mechanic's guide to parts substitution. Although copies of the *Hollander* manuals can sometimes be found in general repair shops, the easiest access to this valuable information source is to buy your own reprinted copies. A set of *Hollander* manuals is quite expensive, so you might suggest that your local truck club purchase one to share among members, or ask a community librarian to add the manuals to the library's automotive section.

Deciding Where to Begin
Rebuilding

If you have decided to take the rebuilding approach, you will keep disassembly to a minimum. For example, a must-do job on any older vehicle is inspecting and overhauling the brakes. If you're working through a brake system for the first time, take only one side apart at a time. That way, if you get stuck putting a front or rear brake back together, and the instructions in the service manual don't make themselves clear, you can always pull off the drum on the opposite side of the truck and use that set of brakes for a reassembly guide—provided, of course, that someone hasn't cobbled a previous repair.

When doing sheet metal repair, it is sometimes advisable to do some disassembly to make the job easier. If you are installing patch panels in the cab, for example, you might also want to remove the doors. *Always* remove the gas tank before doing any welding in its vicinity. You'll need to remove the pickup box to get to the back of the cab and to replace the box's cross braces and flooring. Major mechanical work, like rebuilding the engine, also requires that the mechanical assemblies be pulled out of the truck.

The basic guideline here is to take apart only one or two assemblies at the same time. While the engine is out of the truck for rebuilding, it makes sense to overhaul the brakes and replace the wiring. Overall, though, you will keep the truck as together as possible so that you can continue to use and enjoy it while improving its appearance and operation.

Whenever you take anything apart, be sure to save all nuts and bolts—have empty coffee cans or plastic buckets handy to hold the hardware—and place all parts where they won't get bumped, damaged, or misplaced. As much as possible, have replacement parts on hand *before* starting the repair. This way, your truck won't be laid up while a supplier back orders out-of-stock parts.

Restoring

If you are following the restoration sequence, you will remove the bed, cab, front clip (fenders, hood, and grille assembly), and running boards (on trucks so equipped) to expose the chassis. As you remove the truck's sheet metal, avoid breaking or cutting the bolts, as much as possible. To keep rusty bolts from seizing on the nut and breaking, soak the bolts and nuts well with penetrating oil for several days before attempting to turn them loose. If a nut still refuses to turn, apply heat with a torch. It's important to save the old bolts because new hardware typically does not match the original bolt heads and nuts in thickness. Even though you are careful, some of the bolts will break and others will be disfigured by rust. To replace these, you will need to remove additional bolts from a parts truck or a truck carcass in a scrap yard.

When you have removed all body parts, inspect them closely for damage that was formerly hidden. Places to look are underneath the cab, where you can now check the condition of the support braces. Also check for rusted-out areas on the inside bottom of the doors and in similar spots that are easier to see now that the cab can be moved around.

All sheet metal parts should be stored where they won't be bumped into or damaged. Storage should be inside to prevent further deterioration.

With the chassis laid bare, you can remove the

For rebuilding, the truck will be left as intact as possible. Removing the hood, front fenders, and doors gives access to the cab interior and front mechanical assemblies.

engine and drivetrain. You will keep these assemblies intact for now. This done, you can strip off the front end, steering, and suspension—leaving just the bare frame. More specific instructions for removing these assemblies are given in the chapters that deal with overhauling the engine, front end, and other major components.

Once you have the truck stripped to the bare frame, you're at the starting point of the restoration process. Just as the frame was the platform on which the truck came into being as it moved along the assembly line many years ago, now the cleaned, derusted, and repainted frame will be the centerpiece around which the restored truck comes together. Of course, various assemblies such as the engine will be rebuilt separately. But the truck won't begin to take form until you have refinished the frame.

One tip that is very important to a successful restoration is not to throw anything away until the truck is finished and you are absolutely sure the old part is of no use—even as a pattern. For example, that old, ratty cardboard headliner may look useless when you pull it off the truck, but it may serve as a helpful guide when installing the replacement headliner several months later.

To store small parts, gather a collection of coffee cans or cut the tops off plastic milk bottles. These containers can hold nuts, bolts, and miscellaneous other small items. You can write the contents of coffee cans on strips of masking tape applied to the outside. The milk cartons can be labeled with a permanent marker—"Left Front Fender Bolts and Washers," for example.

As you take your truck apart, make it a practice to label everything. This can be done with tags and masking tape. Then place all the items in well-organized storage. Otherwise, when you start reassembly, you'll waste lots of time looking for misplaced items.

Working Therapeutically, Not in a Frenzy

Working on your truck can be a great escape from the pressures of life if you follow a few common sense guidelines.

Keep Your Work in Perspective

The first guideline is not to let the truck become

Swap meets, usually held in conjunction with car and truck shows, are ideal shopping places for parts for your vintage Ford truck. There you can often find a mix of used, new, and reproduction parts.

the driving compulsion of your life. If this happens, then daily life will become an escape from working on your truck, not the other way around. The simplest way to turn working on your truck from a relaxing pastime into a compulsion is to set deadlines on your work. Repair or restoration of an older vehicle never agrees with deadlines. Sometimes jobs will go more smoothly than expected, but not often. More commonly, the unexpected will throw your work off pace. The most common schedule saboteurs are missing or incorrect parts, but setbacks can also arise that seem providentially placed to test your patience. Further, these holdups usually seem to strike in the final steps just when you're nearing completion. That's when it's important to be able to step back, put away your tools, and admire the progress you've made. When you come back to the truck, the next day or the next week, you'll have thought through ways to surmount the obstacle and your work can proceed calmly toward its goal.

Break Your Work into Manageable Chunks

The second guideline to making the time you spend working on your truck enjoyable is to take the jobs in manageable chunks. Even a frame-up restoration should be handled one job at a time. If the current stage of work is sandblasting, then give your attention to that step and deal with concerns about metal repair later. If while working on larger projects—such as rebuilding the drivetrain, repairing the metal, or refinishing—you sense frustration or impatience setting in, it's a good idea to break off and concentrate on a smaller, more manageable job that you can complete in one or two evenings or a weekend. That way, you'll have the satisfaction of seeing something through to completion and you'll return to the larger project with renewed confidence in achieving your goal.

Work with Another Person

It also helps to work with a partner or friend. Besides getting more than twice as much accomplished, working with someone else has the benefit that when you get discouraged, your partner may be able to provide the encouragement to move ahead—and vice versa. Details like getting the doors or hood to make a perfect fit may take a buddy's saying, "Let's give that one more try."

Alternate Tasks

Another way to avoid old-truck burnout is to alternate between smaller jobs and larger projects. If you live in a northern climate, winters may be a "dead time" as far as working on your truck is concerned. But you can keep the project moving ahead by rebuilding smaller assemblies like the starter, generator, or carburetor inside the house. You can also take this time to send instruments out for rebuilding, or have the chrome plating done.

That way, you'll feel as if your truck is progressing, even if you aren't standing, toes numb, on bone-chilling concrete, gripping an icy wrench.

Looking Ahead

The chapters that follow describe the restoration and rebuilding steps that are typically performed on an older truck. Each is presented as a do-it-yourself process that is easily within the capability of anyone with the do-it-yourself spirit and a willingness to learn. Where special tools are required, these are mentioned with the process. If a more substantial tool investment is required—as is the case with sandblasting—alternative sources to getting the job done are suggested.

One last thought: You don't need to wait until your truck is all finished to show it off to your family and friends. Some restorers have even been known to haul their truck's chassis to shows. Spectators will be amused, and some will later remember the time when your truck rolled in as just a black frame and running gear. Offbeat moves like this are a good way to keep from taking your hobby too seriously. Remember, you're working on that old truck because it's *fun*.

A swap meet vender displays the grille and other body parts from a 1938–39 Ford pickup. When buying parts at a swap meet, check them over carefully. You have little chance of returning them if, when you get them home, you find damage not noticed at first or they don't fit your truck.

Chapter 4

Tool Selection

Several differences typically separate the hobbyist restorer from the restoration professional. These differences lie in the nature of the shop where the work is to be done, the completeness of the tool set used to do the work, and the level of the skills available for performing various mechanical and body repair techniques—as well as familiarity with special tools such as welding and spray painting equipment.

Although hobbyists are often very resourceful in solving their shop space or tool needs—it's not uncommon to hear of vehicles having been restored out-of-doors, though this approach isn't recommended—before starting a restoration or the repair job needed to bring an older truck back to prime, take a close look at your work space and tool set. Compare what you've got available with the recommendations given in this chapter. The chapters that follow will help you acquire the skills needed for the various steps of mechanical and body repair.

Needless to say, you don't have to purchase every tool that might be useful in restoring an older truck—after all, resourcefulness is part of the challenge too. Some jobs, however, such as spray painting, do require the right equipment to do the work. This chapter presents a list of basic tools needed to do the types of repair work typically required on an older vehicle, as well as recommendations for specialized tools that you should either consider purchasing, plan to rent, or work out an arrangement for borrowing.

In the process of restoring or rebuilding an older

A spacious shop like this one is ideally suited to restoration work. You'll find that when you take your truck apart, the amount of shop space needed doubles.

I do my restoration and repair work in this century-old barn built by a Michigan lumber baron.

truck, you will develop and discover skills through trial and error, practice, reading, and getting help and guidance from more experienced friends. Finding those who can guide you through this new territory will be an important step to helping ensure that your old truck will end up the beauty of your dreams rather than a half-completed project that has consumed your energies and finances with nothing in return. You'll find experts who can help you develop the skills needed to rework an older truck at places like the parts counters of auto parts stores, general repair shops, in adult education auto repair courses, and through involvement in a local old-car or -truck club.

As you're sure to discover along the way, many difficult or seemingly impossible jobs are made easy by knowing the right approach. You'll learn some of these tips in this book; others you will pick up from the "experts" you have found to help you when you get stuck. If enthusiasm transcends your mechanical and body repair experience, you may find it beneficial to enroll in adult education classes in auto mechanics, welding, or bodywork at a nearby skill center, vocational high school, or technical college. Often the instructor will let you work on whatever part of your truck relates to the skills being taught—in a welding class, you might be able to practice by installing patch panels, for example. Where this is the case, for the small outlay of a tuition fee, you receive not only valuable instruction, but the use of expensive tools and warm, dry, well-lighted shop space as well.

One basic skill for working on any older vehicle is knowing how to light and handle a gas welding outfit. This chapter explains the basics of gas welding, an essential skill for heating, cutting, and joining metal. As you take your truck apart, a "gas wrench," as mechanics often call a gas welding outfit, will many times seem to be an indispensable tool. Welding as a repair process is discussed in more detail in chapter 10.

Shop Needs

The space needed to store and work on an old truck will grow as the truck is taken apart. For a while, you can keep the truck's spread to a minimum by storing dismantled parts in the bed, but sooner or later, that comes off too. Then you have the spares and replacements, which sometimes include a parts truck or two, that need to be housed—or at least parked. And you'll really kick yourself if you toss out rusted or damaged original parts from your truck, only to find that you need them to check a fit or paint scheme after they're gone. The best policy is to save *everything* until the restoration is finished. All this, plus room to work on the truck, amounts to more shop space than many hobbyists have available. What's to be done?

The first step is to realize that shop space is

The mainstay of any shop is a sturdy workbench. Mine often serves as a catchall storage area.

One way to keep clutter off the workbench is to provide shelving and storage space for small hardware items, lubricants, and other odds and ends or parts.

Painting supplies should have their own storage area. A better arrangement than placing them on open shelves is to store paints and solvents in locked cabinets.

likely to be a problem. If you have a two-car garage, you can get by, but the daily driver or drivers may be parked outside for a long while. If you have less than a two-car garage, you might consider putting up temporary storage for the course of the restoration. And other options can be explored.

For my restoration and storage needs, I lease a nearby barn—once the carriage building to the town's former mansion, which is now a fraternity house. The barn provides space to store my collector vehicles and is an adequate, if not altogether suitable, facility in which to do mechanical work. The building is not heated, so I have to schedule projects for warm weather.

An alternative that works for several collectors is to work with others who have the facilities you don't. Members of the Light Commercial Vehicle Association's Virginia Blue Ridge Chapter take turns bringing their trucks into the "club house"—a member's three-car-plus garage—for specific restoration projects. Another collector does restoration projects in a friend's shop, with the understanding that he will assist on the friend's projects. Undoubtedly hundreds of stories recount how resourceful restorers have solved their garaging problems. The challenge for you is solving yours.

Besides the physical space in which to park the truck, store the parts you remove and others you acquire along the way, and work on the parts you

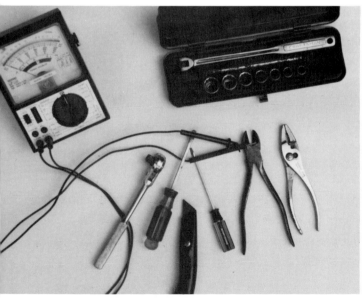

Among the items found in a basic tool set are a socket set with a ratchet handle, pliers, assorted screwdrivers, and a utility knife. The volt-ohm meter, also included here, is a handy tool for electrical troubleshooting.

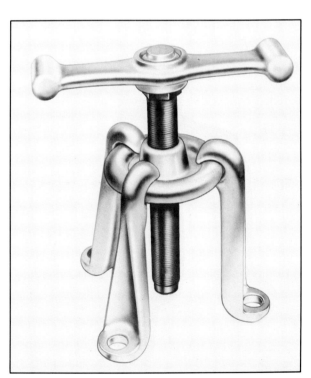

A brake drum puller is needed to remove the rear brake drums for half-ton Ford trucks of Model T vintage to the late forties. These drums fit onto the tapered end of the rear axles and require a puller to slide loose.

have dismantled, a restoration shop also needs some special facilities. To remove the engine, either you will need a strong overhead support beam from which to suspend a come-along or chain fall, or you will need to rent a cherry picker engine hoist. For spray painting, you should install a high-volume fan in a window or vent hole to exhaust paint fumes and dust. If you plan to do your own finish painting, you will need to rig a spray booth; this can be done by putting up "walls" of sheet plastic and installing a ventilation system. For welding, some portion of the floor should be made of concrete or another noncombustible material. Parts storage will take less space and be more organized if you set up shelving. Paint and chemicals should be kept in locked cabinets to prevent their access by children. You will need a sturdy workbench, storage places for tools, 220-volt service for an arc welder and air compressor, good lighting, and a heat source if you live in a seasonal climate and plan to work in cool weather.

If you are in the position to construct a shop for restoration purposes, then you have the opportunity of considering helpful extras like running water with a sink for cleaning up after work and a floor drain for washing your vehicles. A set of permanent ramps or a pit for servicing your truck and doing repairs is handy.

You'll also find it helpful to run galvanized pipe for air compressor lines, with outlets in various locations in the shop. Running pipe for the main compressor lines has at least three advantages over uncoiling a length of air compressor hose each time you use air tools: first you avoid the nuisance of having to uncoil and coil the hose; second you won't lose air supply through restrictive hose connections; and third the pipe will assist with moisture condensation, helping prevent water from passing into sandblasting or spray painting equipment and air tools.

Those who plan to do their own sandblasting are advised to construct a three-sided pen outside the building to catch the dust and conserve the sand. If you set up a degreasing bath of the type described in chapter 5, then you should also construct an enclosed area, either inside or outside the shop, for the alkaline bath. It is crucial that you take every measure to prevent children—your own or those in the neighborhood—from gaining access to harmful chemicals that you may be using in the restoration process. You will also need to work out an environmentally sound means for disposing of the alkaline degreasing solution and other chemicals that if simply dumped out would leach into the soil and contaminate ground water.

Basic Tools

Having a vehicle professionally restored can be a very expensive experience. The only way I know that you can avoid a professional shop's $30 to $50 hourly fee is to restore or repair the vehicle yourself. The repair or restoration will probably take you a little longer than if you hired out the work, and along the way, you're likely to encounter a few unpleasant frustrations. But when the job is done, the feeling of accomplishment will be very rewarding and the money saved will be substantial.

Doing your own restoration and repair work typically requires some investment in tools. Trying to restore or rebuild an older truck using an inadequate tool set is bound to produce frustration and damaged

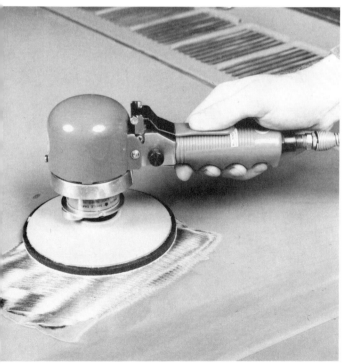

Once you have an air compressor, you can invest in a variety of air-driven tools like this dual-action sander. You will find this sander indispensable for stripping off paint and preparing metal for painting. The Eastwood Company

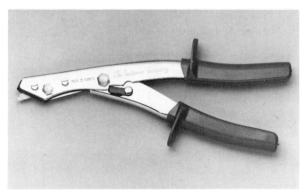

Other helpful tools include a metal nibbler, used to cut out rusted metal and make repair patches. The Eastwood Company

or incorrectly repaired parts. It isn't necessary to own every tool in the professional's shop, but the home handy person's tool set won't be adequate either. The tool set described below will get you through most restoration and repair jobs. Remember that money spent on tools can be considered a lifetime investment, and you can reduce some of the tool costs either by renting from a rent-all store or by teaming with friends or club members to buy more expensive tools on a cooperative, shared basis.

Socket Wrenches

The single most-used tool for mechanical repair is the 3/8in drive ratchet and sockets. The 3/8in drive ratchet is big enough to tackle most jobs on a light truck, and it is still small enough to get into tight spots. The socket set should contain ten sockets ranging from 3/8in to 15/16in in increments of 1/16in. Sockets generally come as either twelve-point or six-point. Unless you have a specific need for twelve-point sockets, I recommend the six-point ones. The six-point sockets have more surface area to grab a nut or bolt, making it easier to remove the fastener without "slipping" or damaging it. For this reason, they are far more effective with rusted bolt heads and nuts than are twelve-point sockets. It is also a good idea to have a swivel coupling and a range of extension bars.

Adding to this basic set, a 1/4in drive ratchet and socket set is handy for really small jobs, and a 1/2in drive ratchet and socket set is needed for larger jobs. On a light truck, you can get by without the 1/4in set, but you'll need the 1/2in drive bar and sockets to loosen larger, rust-frozen nuts and bolts.

Combination Wrenches

Combination wrenches have one open end and one box end—hence the name *combination*. This style of wrench is handy for hard-to-reach locations where if you can't turn the bolt or nut with the box end, you can usually get a bite with the open end. A set of combination wrenches should range from 3/8in to 1in in increments of 1/16in.

Chisels and Punches

Chisels will be used to remove rivets and corroded nuts and bolt heads that are too rusted or have rounded over and so will not hold a wrench. A set of four chisels ranging in width from 3/8in to 3/4in should cover all jobs.

The punch set should include starting and drift punches, which are used to drive out rivets and roll pins; an aligning punch for lining up holes; and a center punch, which is used to make an identifying mark for centering a hole. A brass drift is also a valuable tool and is used for replacing bearings or other easily damaged parts.

Pliers

A variety of pliers are needed for jobs that require pinching, removing, cutting, squeezing, and holding small parts. Pliers that should be included in the basic tool set are standard slip-joint pliers, 9in channel locks, needle-nose pliers, and diagonal cutters.

Hammers

Two ball-peen hammers are needed: a 12-ounce (oz) for light jobs and a 24oz or 32oz for big jobs. A plastic-tipped or brass hammer is critical for any job where it is possible to scratch or damage the part with a steel ball-peen hammer.

Screwdrivers and Miscellaneous Hand Tools

The basic tool set should contain a variety of screwdrivers, ranging in shank length and blade size, and with Phillips as well as slotted heads. Miscellaneous hand tools include a hacksaw and a selection of flat and round files. If you're building your tool inventory from scratch, you can buy lifetime-guarantee Craftsman tool sets containing most of these items, at a relatively low cost from Sears.

Jack Stands and a Jack

To support the truck chassis while you are doing mechanical work like rebuilding the front end, professional-grade jack stands are a must. In

A sandblaster is almost essential for restoration work, since sandblasting is the most versatile way to remove paint and rust.

selecting the stands, avoid the light-duty versions sold in discount marts. The heavier-duty stands have a ratchet release for the load head and thicker-gauge steel legs and braces. Never support the truck on a stack of blocks. Jack stands are relatively inexpensive items and should be selected as though your life depended on the quality of their design and construction—which it does.

Although a regular platform-style hydraulic jack will be adequate for raising the truck off the floor, you will find that a floor jack with rollers is a much more convenient tool to use. It also has the advantage of letting you wheel the truck around on the shop floor, sometimes enabling you to move the vehicle into an otherwise inaccessible space.

Specialty Tools

In the process of your truck's restoration or rebuilding, specialty tools may be needed to perform certain jobs or to make those jobs easier. Specific tools for specialized jobs are identified in the various chapters of this text. The specialty tools listed below should be considered "musts" if you are planning to do a frame-up restoration.

Brake Tools

Nearly all older vehicles need a thorough brake overhaul. The tools needed to completely redo a hydraulic brake system include a tubing bender and flaring tool for creating replacement brake lines of the right length and shape to match the original, and a tubing cutter and brake spring spreaders.

Pullers and Engine Overhaul Tools

Pullers are needed to remove the steering wheel; to remove the crankshaft pulley, if you decide to rebuild the engine; and to take apart other mechanical assemblies. The types of pullers needed for these jobs are shown in the Ford service manual for your truck.

An engine overhaul has its own tool list, including valve spring and piston ring compressors. A gasket scraper and wire brush are handy items for parts cleanup.

Often needed with a mechanical overhaul and critical to engine work are measuring tools. For basic measurements, I suggest a linear caliper. This is not as exact as a micrometer, but it can be quite accurate to three decimal places and is more affordable.

Torque Wrench

A torque wrench, which is used with 1/2in drive sockets, allows you to tighten nuts and bolts to the proper torque specifications. It should be used to tighten everything from head bolts to lug nuts, and will prevent stripped threads due to overtightening or damage that results from parts coming loose owing to undertightening.

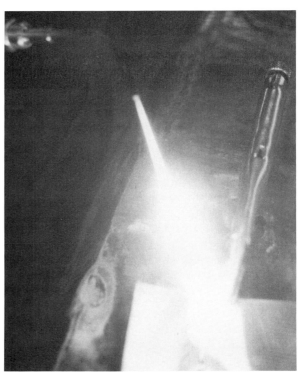

The first step in gas welding is lighting the torch. To do this, turn on only the fuel gas valve and light it with a torch lighter. Do not use a butane cigarette lighter to ignite a gas welding torch; the lighter could explode, causing serious injury or death. Once the fuel gas flame has been lit, the oxygen valve at the torch handle is opened and adjusted until the correct shape and heat intensity of flame are achieved.

Gas Welder and Air Compressor

Although they require a larger investment than any of the tools mentioned previously, a welding torch and air compressor are essential for any serious restoration or repair work on an older vehicle. If you have to stagger the purchase of these two items, buy the torch first. You will use it during disassembly. The air compressor can be used for sandblasting and painting and is an excellent power source for a variety of air tools.

When you buy a gas welding outfit, you typically purchase the gauges, hoses, and torch, and rent the tanks. The oxygen and fuel gas tanks come in a variety of sizes and are best moved around the shop on a cart—also purchased.

As an alternative to a full-sized welding outfit, you can purchase small, portable gas units like the Toteweld by Arco. The Toteweld has the same capability as the bigger setups, but a more limited capacity.

When selecting a gas welding outfit, it is essential to pick one that uses oxygen cylinders; the cheap oxygen pellet units have neither the capacity nor the durability for the heat applications you will encounter while disassembling your truck. For the fuel gas, you can use either acetylene or methylacetylene propradiene (MAPP); propane does not produce enough heat.

In selecting an air compressor, consider the horsepower rating of the motor and the size of the air storage tank. Hobby-size portable air compressors with motor ratings of 3hp or 4hp will be adequate if also coupled with a 30-gallon (gal) or larger air tank. For sandblasting and extensive use of air tools, you are better off to consider a stationary air compressor with a 5hp or stronger motor and a 40gal or larger air tank.

Introduction to Gas Welding

One of the most useful tools for restoring or rebuilding an old truck is a gas welding outfit. Mechanics often refer to their welding torch as a "gas wrench," and for good reason. Rust-frozen exhaust manifold studs, kingpins, and chassis bolts are virtually impossible to remove without heat. If you do manage to turn the fastener, it will more than likely break off—and in the case of a manifold stud, this means drilling out the bolt and retapping the hole, a very time-consuming and frustrating task. The only way to remove rust-frozen kingpins without heat may be to disassemble the axle as much as possible and take it to a machine shop, where the pins can be forced out with a hydraulic press. Using a gas welding torch to heat the axle ends is a much easier and quicker approach. Of course, a gas welding outfit can be used for repairing metal as well. Once you purchase a gas welding setup, you will wonder how you worked on older vehicles—and newer ones too, for that matter—without it.

What you purchase is the set of gauges, hoses, and welding and cutting torches, plus several sizes of torch tips. The tanks—one for oxygen and the other for the fuel gas—are typically rented. This equipment, plus a cart to move the welding outfit around in your shop, is available in welding shops, auto supply stores, and tool retailers like Sears.

Most gas welding outfits use acetylene as the fuel gas. When burned in pure oxygen—from the oxygen tank—acetylene yields temperatures in the 1,800-degree (Fahrenheit) range required to heat metal to a cherry red state. MAPP is an alternative fuel gas that has several advantages for the hobbyist. First it rates superior to acetylene in safety and ease of handling. Second, since it is distributed in liquid form, a MAPP canister provides more fuel per pound than an equivalent weight of acetylene. Third MAPP has nearly as much heat value as acetylene, making it effective for both welding and cutting. And fourth, MAPP has a distinctive odor that you will recognize right away if you accidentally leave the shutoff valves open.

Safety

When gas welding, it is essential that you carefully follow safety rules. If you enroll in a welding class at a skill center or technical college, these should be taught at the start of the class. If you learn to use a gas welding outfit from a friend, or by reading a book on welding and practicing, you may overlook these important procedures. The principal safety rules to follow are listed below.

<u>Never weld near flammable vapors (a gas tank).</u>

Never place the tanks in a position where they might fall over; a tank with a broken valve can be lethal.

Never run over the welding hoses with the welding cart or damage the hoses in any other way. Keep the hoses coiled up when not in use.

Never oil the regulators, hoses, torches, or fittings. The combination of oil and oxygen can produce a deadly explosion.

Never use the oxygen jet to blow away dust.

Never allow the oxygen jet to strike oily or greasy surfaces.

Always wear welding goggles when looking at the flame.

Always wear welding gloves and protective clothing.

Do not carry a butane lighter in a pants or shirt pocket when welding. A spark from the welding operation could explode the lighter with deadly consequences.

Gas Welding Techniques

Although sheet metal welding takes some skill and practice, very little skill is needed to use a gas welding torch to heat rust-frozen parts. Since you are not going to cut the metal, you do not need to use a cutting torch. Rather you will use a welding torch with a medium to large tip.

If you have purchased a gas welding outfit but have not had instruction in how to use it, then learning how to set the dials on the tanks and light the torch should be your first step. Instruction in both gas and arc welding is available at a low cost from technical colleges and adult education programs sponsored by skill centers or vocational high schools. The classes are typically held in the evenings and last for ten or fifteen weeks. Enrolling in a welding class will thoroughly acquaint you with safety principles as well as instruct you in the theory of welding technology and provide sufficient practice to make you a moderately skilled welder.

The steps for using a gas welding torch to heat rust-frozen bolts that are given below are not intended to replace formal instruction. Instead they are intended as a quick reminder or review.

Step One

Before heating any rust-frozen bolts or other part to be removed from a vehicle, make sure that you will not be directing the torch toward any flammable material—the gas tank, gas line, undercoating, and so forth.

Step Two

In preparation for lighting the welding torch, open the regulator valves on both the oxygen and fuel tanks. Be sure that both valves on the torch handle are closed before opening the regulator valves. Now open the valve on the oxygen cylinder (always painted green) by turning it counterclockwise. Turn the valve slowly until the regulator pressure gauge reaches its maximum reading. Then turn the valve all the way open. Next open the fuel tank valve a quarter turn. This is so that the fuel cylinder can be shut off quickly in an emergency.

Step Three

Adjust the oxygen pressure at the torch by opening the torch valve on the oxygen line (also green) and adjusting the oxygen regulator so that the low-pressure gauge, which registers the flow of oxygen to the torch, reads between 8 pounds per square inch (psi) and 20psi. When this setting is achieved, close the torch valve on the oxygen line.

Next adjust the fuel pressure by repeating this procedure with the torch valve on the fuel line. Set the fuel regulator so that the pressure reading is between 8psi and 9psi; this is the pressure setting for a medium to large tip; a small to medium tip would require a pressure setting of 4psi to 5psi.

Step Four

Before lighting the torch, it is important to purge the lines. This is done by briefly opening both torch valves. As soon as you hear the hiss of gas escaping from the torch, close the valves.

Step Five

Now you will light the torch. Begin this procedure by opening the fuel valve slightly—approximately one-half turn. As soon as the fuel valve is opened, strike a spark at the tip with a torch lighter. Do not use matches, and *never* use a cigarette lighter to ignite a welding torch.

If you have opened the fuel valve too far, the flame will ignite, then blow out. If this happens, close the fuel valve slightly and strike a spark at the tip again. A yellowish flame will lick out of the torch tip.

Step Six

Open the oxygen valve at the torch slowly and gently. As you do so, the flame will change in shape and color. If you open the oxygen valve too far or too fast, the flame will blow out with a pop. If this happens, don't be alarmed. Just close the oxygen valve and relight the flame.

Step Seven

Once the flame is burning in the oxygen supplied through the torch—rather than in the air—the oxygen and fuel gas mixture needs to be adjusted for maximum heat. This is done by turning both valves until the correct proportions of fuel and oxygen are emerging from the tip. To achieve this setting, put on your welder's goggles and slowly open the oxygen valve. You will see the flame separate into three distinct parts: a small, light-colored cone at the tip, surrounded by a darker-colored cone, and the flame's outer halo. A flame with a larger outer cone has too much fuel and too little oxygen. Continue to open the oxygen valve until the outer cone disappears and the inner cone has a smooth, round shape.

When you have adjusted the flame to this configuration, you may need to increase the flow of both fuel and oxygen, in equal proportions, to achieve a hotter flame. As you increase the flame's strength, you will probably also need to readjust the oxygen and fuel settings to bring the flame back to the correct cone shape.

Step Eight

When heating metal to loosen rust-frozen bolts or kingpins, always apply the heat to the area around the fastener—never apply it to the bolt or pin itself. Heat enables you to loosen a rust-frozen bolt by expanding the metal into which the bolt is threaded, thereby breaking the rust's grip. If possible, play the flame all the way around the bolt, keeping the torch moving, until the entire area begins to glow a cherry red. If you are heating a casting, like an exhaust manifold, and fail to keep the torch moving, heat will build in one spot and may crack the part.

Step Nine

When the metal begins to show a reddish glow, you can turn off the torch, grip the bolt with a socket or combination wrench, and turn it loose. The bolt should turn fairly easily. If it doesn't, apply more heat—making sure to direct the torch toward the surrounding metal.

Step Ten

After removing the bolt, *do not* touch it with your bare hands and do not let it drop or lie on a wooden floor. The bolt will have reached a temperature of several hundred degrees Fahrenheit from the heat transferred through the adjoining metal, and will take quite a while to cool.

Step Eleven

When you are finished using the welding torch, close the valves on both the oxygen and fuel tanks, then purge the lines by opening both valves at the torch. You will hear a momentary hiss as the gas in the lines escapes. Now close the valves at the torch, and roll up the hose and hang it on the tanks so that it does not drape on the floor. Then roll the welding cart to a space in the shop where it will be out of the way.

A gas welding outfit is one of the handiest tools in an old truck restorer or rebuilder's shop. Not only does it make disassembling parts far easier, but it also prevents parts from being damaged beyond reuse during the disassembly process. Later chapters will show how the gas wrench can also be used to repair rusted or damaged sheet metal.

Spray Painting Gun and Air Tools

Even if you decide to have a professional painter apply the finish coat, you will save a great deal of money—and transportation hassle—if you apply the primer coating. For any automotive painting, a spray painting gun with an external air source is essential. The quality of the gun determines, to a large extent, the quality of the paint application.

Along with the spray gun, you will need a professional-grade painting mask, a moisture filter at the compressor outlet, and quick-connect couplings on the gun and air hose. Chapter 11 provides a more detailed discussion of spray painting equipment and describes the newer-technology, high-volume, low-pressure (HVLP) spray painting equipment that should be considered as a more environmentally sound alternative to high-pressure spray painting.

Air tools save time with both mechanical work and bodywork. Among the popular air tools are an impact wrench; a die grinder; a drill; a cutoff tool; and inline, rotary, and dual-action sanders. These tools need a large volume of air to operate efficiently; hence the concern for large air storage and recovery capability.

Sandblaster

As discussed in chapter 5, two types of sandblasters are available: siphon and pressurized. Siphon sandblasters are inexpensive and work well for smaller jobs. If you purchase an air compressor, you will most likely want to add a sandblaster to your tool inventory. You will find that owning one

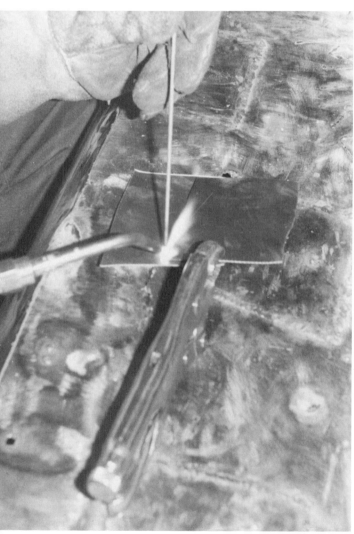

An overlap joint is one method used for welding in patch panels. Unless you are an experienced welder, you should practice with small pieces of metal scrap before setting out to weld patch panels onto your truck. Note that the two pieces of metal are clamped tightly together with ViseGrips.

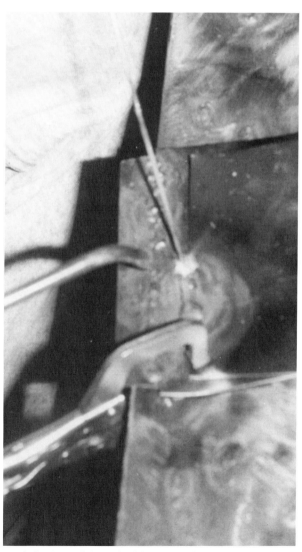

To help control heat buildup—and keep from distorting the metal—welds are spaced across the seam approximately 1in apart. On a larger panel, the welds would be made in an alternating pattern, first at one end of the seam, then at the other.

boosts your popularity with fellow old-truck restorers.

Brightwork Tools

The restoration of a vintage vehicle often calls for bodywork and mechanical repair work that is quite different than that done on newer vehicles. One example of this is the need to straighten and polish bright metal trim. On a newer vehicle, you would simply buy replacement trim pieces, but for an older Ford pickup, these are often not available.

To straighten bright metal trim, special hammers are available. The trim is polished and buffed using special buffing wheels and compounds. A wide assortment of specialty tools for this type of restoration work is available from The Eastwood Company (see "Suppliers" in the Appendix). Ordering Eastwood's catalog will familiarize you with tools that apply for specific applications. When specialty tools are needed, or advised, for the various repair or restoration procedures described in this text, they will be listed in the repair sequence.

Tool Storage

As your tool collection grows, you will need a storage system that makes the tools easily available. The three basic choices for tool storage are a toolbox, a roller cabinet, and a combination of both.

The toolbox must have easily working drawers, a functional lock, and enough room so that the tools don't lie on top of each other. After that, it is personal preference.

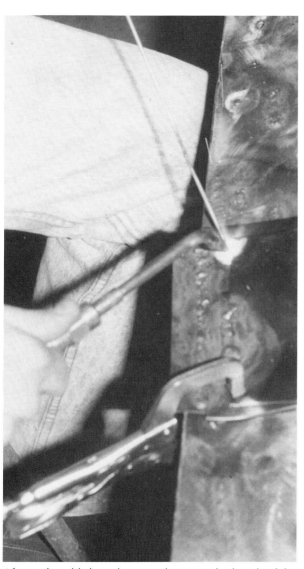

After tack welds have been made across the length of the seam, the gaps between them can be filled in. Here, too, the alternating pattern would be used—welding a gap at one end of the panel, then at the other.

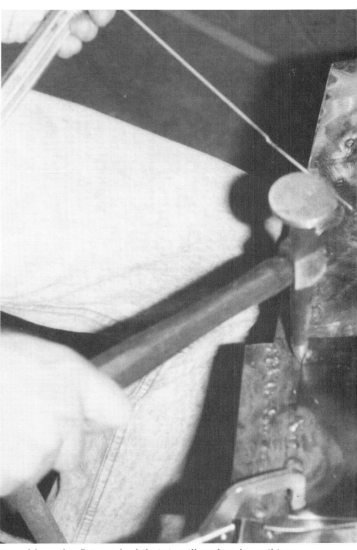

A weld can be flattened while it is still molten by striking it with a hammer. A variation of this technique, called hammer welding, is used by experienced bodyworkers to join metal with seams so smooth that little or no filler is needed.

When heat warpage is not a problem, as in repairing cracks on thicker-gauge metal like that on this cowl band from a Model A Ford roadster pickup, the welder can safely lay down a continuous bead of filler metal.

The basic tool set should fit into a toolbox. As you add to your tool collection, you will find a roller cabinet better suited to your storage needs. Beware of low-quality roller cabinets, often seen in discount marts, that are undersized, are flimsily built, and have such small rollers that they do not move around easily. Quality roller cabinets and accompanying toolboxes that are designed to sit on the top of the cabinet are available from Sears as well as professional tool suppliers like Snap-on Tools Corporation.

Quality versus Price

Tools can be found in three categories: bad, good, and excellent. Bad tools are the cheapest, in both cost and quality. Examples are the non–name brand tools found in discount marts. These really aren't worth bringing home. I bought one of the cheap screwdriver sets once and chipped the blade on the first screw I attempted to turn.

Examples of good tools are the Sears Craftsman line. Craftsman tools come with a lifetime guarantee, have a high standard of quality, and sell at a reasonable price. A plus is that most Sears stores will replace a broken tool on the spot with no questions asked.

Excellent tools are Snap-On, Mac, or other professional brands. These are the highest-quality tools made—and the most expensive. They not only have a look of quality, but feel different in your hand—like a well-broken-in baseball glove. These tools also come with a lifetime guarantee, and in most cases will outperform any other make, including Craftsman. One measure of tool performance is the tool's "fit" on a rusted bolt. A Snap-on dealer once had me try one of his wrenches on a rusted bolt that my Craftsman wrench would simply slip off. The Snap-on wrench not only took a firm grip, but turned the bolt loose.

The quality-to-price ratio of Craftsman tools is hard to beat. If you are operating from a budget, these are the best way to go. But if you truly enjoy mechanical work and expect to work on vehicles for years to come, and can afford them, then Snap-on, Mac, or other professional tools are what you want. Their quality look will increase your pride in your work, and you will find that once you have developed the "feel" of working with high-quality tools, a lesser-quality tool will seem like just a chunk of steel in your hand.

Chapter 5

Stripping and Derusting

As you take your truck apart, you will encounter years of caked-on grease and grime on mechanical parts, and probably some rust on the body metal. Not only does removing the grease buildup make rebuilding mechanical assemblies like the engine and front end easier, but a thorough cleaning is also important if your truck is to have that "fresh-from-the-showroom" look. Welding can't be done effectively on rusted panels, and paint won't stick to rust, so removing the paint and derusting the body metal are necessary preparation steps before bodywork and refinishing.

Cleaning, stripping, and derusting are three of the biggest challenges facing any old-truck rebuilder or restorer. Your thoroughness with these steps will determine the outcome of subsequent rust repair and refinishing—as well as the truck's overall appearance when you're done. These steps actually

The first step in the restoration process is removing years of accumulated grease and grime on the mechanical parts, and rust and paint coatings on the body metal.

Those desiring to restore a vintage Ford truck for driving rather than showing will want to note the IFS grafted onto this early-fifties Ford panel truck owned by Keith Ashley.

consist of several different processes that can be done either separately or, in some cases, in combination.

These processes are degreasing, sandblasting, chemical paint stripping, chemical metal cleaning, acid derusting, and rust neutralizing. This chapter describes various methods for performing these tasks on your truck. If you are taking the restoration approach, cleaning, stripping, and derusting are the first steps after complete disassembly. If you are taking the rebuilding approach, you will clean and strip various mechanical components as you overhaul them. Derusting may occur in stages as you repair and refinish your truck's sheet metal.

As you will discover, cleaning, stripping, and derusting can be time-consuming and it is more efficient to do these processes all at once rather than piecemeal. This means that if you are taking the rebuilding approach, you may find it beneficial to clean, strip, and derust as much of your truck as possible at one time, even though you are not taking the whole truck apart.

Degreasing

Typically an older truck's engine and running gear are coated with grease—often an inch or more thick in places. The body metal and chassis may also be covered with a scale of rust. Several methods can be used for cleaning heavy grease from engine and chassis components. The easiest is to take the chassis—or the truck, if you are using the rebuilding approach—to a shop that does steam cleaning, if such a place exists in your community.

If a steam cleaning service isn't available, you may still have the option of renting a steam generator from a tool rent-all service and cleaning the chassis yourself. Steam jennies, as they're called, are about the size of a large portable air compressor and consist essentially of a pump and boiler that produce a high volume of steam, which is then blasted against the grease-coated chassis parts at great pressure. To help in the cleaning action, detergent is often mixed with the water being converted into steam. Using a steam jenny requires no special training except the caution to wear protective clothing to avoid scalding yourself with the steam. The steam nozzle is simply worked back and forth against the grease-coated chassis parts until all grease and dirt have been washed away. Using a steam jenny, you can clean an entire chassis, including the engine, transmission, and rear axle, in a few hours.

If a steam jenny isn't available from a local tool rent-all, you may find a power washer to be nearly as effective. This tool, which can be connected to a hot-water faucet, also mixes detergent with high-pressure water spray for fast, effective cleaning. A power washer is a tool you may wish to own. It can also be rented from almost any tool rent-all.

Another alternative is to haul your truck's grease-coated mechanical parts to a nearby car wash and use the engine cleaner cycle to degrease them. You will find this approach to be more effective if you scrape off heavy grease buildup first with a putty knife. The car wash method works essentially like the power washer method. The difference is that you bring the parts to the wash location, as opposed to the other way around.

If you are working from a low budget and time is of less concern than expense, you can clean the chassis the old-fashioned way with a putty knife, scrub brush, and solvent. This approach can be very time-consuming, but the only cost may be for a few scrub brushes and several gallons of solvent.

When degreasing engine and chassis parts by hand, you'll save yourself a lot of time by first

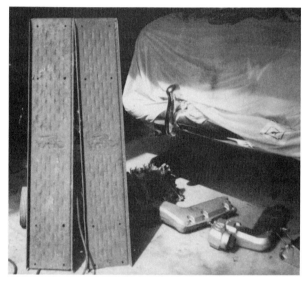

Even well-preserved parts like these Model T running boards need cleaning to remove surface rust.

The degreasing cycle at a coin-operated car wash works well for removing grease and grime from mechanical parts.

scraping off the buildup, using a good grease-cutting solvent like Gunk, and working in warm temperatures. After one or two Gunk applications, the part should wash clean enough for painting.

Before painting, it's important to wash the part with a metal preparation solvent. This is to make sure the metal is clean of all traces of grease and degreasing agent. Gasoline should not be used as a degreasing solvent. For washing parts, use a less flammable fuel like kerosene, or a commercial parts-cleaning solvent. Rubber gloves and long clothing should be worn when working with any solvent. Be sure to protect your eyes by wearing safety glasses or a face shield.

A very effective degreasing bath for cleaning smaller parts can be made by setting up an alkaline solution in a metal pail, discarded oil drum, or even as large a container as a livestock watering tank. The alkaline degreasing solution consists simply of household lye—sold in grocery stores as either lye or Drano—and water in a mixture of approximately two 12oz containers (24oz) of lye per gallon of water. Using a metal container is advised because then heat can be applied to the degreasing bath.

In setting up an alkaline bath, I have used a 55gal oil drum into which I poured about 40gal of water and eight to ten 12oz containers of lye (96oz to 120oz). To heat the degreasing solution, I placed the oil drum on concrete blocks to allow space for building a small wood fire underneath. As an alternative to a wood fire, a propane heater can be used.

When heated to near boiling, the lye solution will remove even heavily caked-on grease coats in just a few minutes. The alkaline solution works equally well removing paint. I used that oil drum degreasing tank to strip and clean wheels, front end assemblies, transmission housings, and engine blocks, as well as other small parts.

Caution: Mechanical assemblies should not be immersed in the alkaline solution unless they are taken apart or will be disassembled and rebuilt later.

Numerous other cautions also apply to the alkaline bath degreasing method. Always wear rubber gloves and a full covering of clothing. Tie lengths of wire to larger parts before dipping them into the bath so that you don't have to fish them out of the solution. Place smaller parts in a wire mesh container. Always wear face protection to prevent the alkaline solution from splashing into your eyes.

It is extremely important to make sure that children and animals cannot reach the alkaline bath. Options are to set up the solution inside your shop and make sure the container is covered and doors are locked when the bath is not in use, or to build a secure enclosure around the bath if it is outside.

You will need to haul the solution to a sealed landfill after the degreasing business is finished.

Although the alkaline degreasing method is very effective, if you question your ability to follow these safety requirements, you should use another approach.

Sandblasting

Like grease, paint and rust can be removed by several methods. The most common and inexpensive method is sandblasting. Other methods use chemicals. Attempting to remove paint and rust by

An alkaline solution made from household lye makes an effective degreasing agent. The container can be plastic or metal.

This collection of early Ford parts has been immersed in the degreasing bucket for several days. The temperature of the solution affects the speed at which it works. Besides cleaning off grease, an alkaline bath will also remove paint and light surface rust.

Sandblasting is the most common and inexpensive method for removing rust and paint. In addition to a sandblaster, you will need a high volume of compressed air. With sheet metal, there is a risk that the high-pressure blast can warp the metal.

If you have no zoning restrictions, or neighbors to complain, sandblasting is best done out-of-doors. A backdrop, like the simple plywood barrier shown here, forms a collecting point for the sand, which can be shoveled back into the sandblaster and recycled. Mike Cavey

sanding, grinding, or using a wire brush is not very effective, since these mechanical methods can abrade the metal and do not successfully remove all traces of rust. The microscopic "seeds" of rust that remain at the bottom of pits will grow, continuing to eat into the metal and blister the paint.

In many communities, you will find commercial sandblasting services listed in the Yellow Pages. These services will strip paint and rust from chassis parts, wheels, bumpers, and sheet metal for prices that are generally reasonable. The main disadvantage to commercial sandblasting is your lack of control over the sandblasting process. Although a commercial sandblaster isn't likely to damage a bumper or frame, high-pressure sandblasting can easily warp sheet metal. If you will be resorting to sandblasting to strip and derust body parts, you need either to make sure the commercial sandblaster agrees to be very careful with these parts to avoid warpage—and then expect to do some straightening—or to sandblast the sheet metal parts yourself, taking great care not to stress the metal.

If you live in a suburban development, sandblasting at home may be contrary to zoning regulations and is sure to upset your neighbors; the process is both noisy and dirty. If you live in a more rural setting where sandblasting will be a nuisance only to you, the equipment investment is reasonable enough that you may well decide to set up your own operation. To do so, you will need an air compressor—preferably a 4hp or larger unit with a minimum of a 30gal air storage capacity. Heavier-duty home-shop air compressors are available from Sears as well as from farm and specialty tool suppliers. An air compressor should be considered a basic shop item, as it is used to power air tools— impact wrenches, sanders, grinders, drills, buffers— and spray painting equipment, as well as a sandblasting outfit.

Most portable air compressors need two modifications before being connected to the sandblaster. The first is to install a moisture separator between the outlet connection on the compressor and the air hose. Typically, the air hose is connected directly to the air tank. This allows moisture that builds up inside the tank, particularly when the compressor is operated on humid days, to travel through the air hose to the sandblaster. Any moisture will clog the sandblaster almost instantly. The moisture separator does what the name says; it removes moisture from the air stream, allowing only dry air to travel down the air hose. This inexpensive device can be purchased from the company selling the air compressor or from almost any tool supplier.

The second modification required for many hobby-sized air compressors is to replace the 1/4in-diameter air hose with a minimum 5/16in–inside

One benefit of sandblasting is that the main ingredient— sand—can often be shoveled free from a sand pit or beach. The sand will need to dry thoroughly before being used, as any dampness will quickly clog the sandblaster's nozzle.

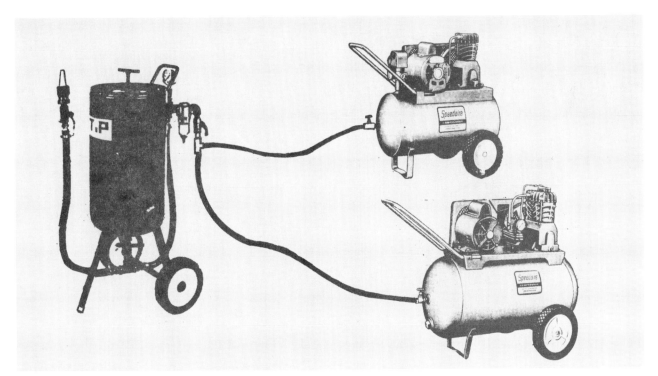

If your air compressor is a portable home-shop unit with limited capacity, you might borrow a friend's and hook the two in tandem.

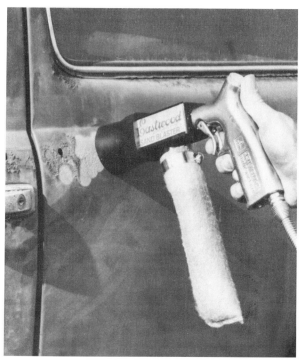

A spot sandblaster is ideal for removing and repairing localized rust. The Eastwood Company

diameter air hose. Keep the hose length to no more than 50ft. Sections of hose should not be spliced together, owing to the air restriction at the connection.

These modifications are also recommended for running air tools and for spray painting.

Sandblasting units are of two types: siphon and pressure feed. Both are available from restoration tool suppliers like TiP Sandblasting, The Eastwood Company, and others. The siphon-style sandblaster is both less expensive and less effective than the pressure feed unit. It draws sand into the air stream by suction. The result is that the blast from the siphon sandblaster is not as strong as that from a pressurized sandblaster, and uses more air—an important issue considering the limited capacity of the air compressors typically found in the hobbyist's shop. This does not mean you should rule out a siphon sandblaster. If you will be sandblasting primarily small parts—either because your truck is largely rust free or because you plan to have the larger parts commercially sandblasted or chemically derusted—the siphon blaster will work fine. I own a siphon unit and have used it to clean my truck's door trim panels, license plate brackets, and other smaller parts. It handles jobs of this size easily. On

As this freshly sandblasted 1950 Ford pickup frame shows, a factory-fresh surface is the starting point of a quality restoration. Chrystal Edgerly

anything larger, however—a wheel, for example—the siphon sandblaster has proved to be inefficient and more trouble than it is worth.

A pressure-style sandblaster does what the name implies; it pressurizes a sealed container and uses that pressure to blow sand out through the nozzle. The result is a more powerful blast and more efficient use of compressed air.

There isn't much art to sandblasting. Basically you point the nozzle at the part and blast off the paint and rust. This process isn't very effective against grease, which is why any grease buildup has to be removed before you sandblast. You will find that blasting at an angle "washes" off the paint and rust and puts less stress on the metal. On large, flat panels like the hood, it is best to sandblast only on reinforced areas and to strip off the remaining paint with paint remover or a dual-action sander. If the panel shows rust in areas that are not sandblasted, a chemical rust remover or neutralizer can be used.

Other sandblasting precautions include the recommendation that you wear full-length clothing with protective gloves (welding gloves work well), a sandblasting hood, and a dust mask. You will see some sandblaster operators who tie off their shirt sleeves and pants cuffs in an attempt to prevent dust from blowing inside their clothing. But dust sifts in anyway through the fabric.

Special coarse-grit sandblasting sand is available from most auto parts stores, but natural sand from a building supplier, beach, or sand pit works nearly as well and is less expensive—often free. The advantage of sandblasting sand is that its coarse grit allows it to be reused, sometimes several times.

If you shovel your own sand from a beach or sand pit, be sure it has dried thoroughly before sandblasting. Nothing produces greater frustration than trying to sandblast with moist sand. The nozzle will clog as soon as you start, and each attempt to unclog the blockage will be followed by reclogging. If the sand is damp when you are ready to start blasting, spread a tarp in a sunny area of your driveway and cover it with a thin layer of sand. When the sand turns white, it is dry and ready to be used.

If you decide to blast with sand you've gathered yourself, you will need to strain it through a window screen or similar fine mesh before blasting. If you don't, larger pebbles and debris will clog the sandblaster's nozzle.

Sandblasting at home with a 3hp to 4hp air compressor really puts a strain on the air supply. If you find the air compressor you own to be inadequate for the amount of sandblasting you are doing, one solution is to borrow a friend's portable compressor or rent a portable unit from a tool rent-all, and power the sandblaster with both compressors. This can be done by connecting the air

Fresh metal needs to be treated to prevent rusting, which will occur rapidly if any moisture is present. For short-term rust protection, and to help paint adhesion, sandblasted parts should be treated with a dilute phosphoric acid solution. The Eastwood Company

The safest, most effective method for removing paint, rust, and other coatings is reverse electrolysis. This metal-cleaning process is available only from commercial chemical stripping services, many of which operate as franchises in the RediStrip chain. Since this process is nonabrasive and does not use acids, it cannot distort or weaken the metal.

lines from the two compressors with a Y-fitting. Borrowing a friend's compressor might present an opportunity for you both to work together sandblasting parts for your vehicles. Sandblasting with two people will be a faster operation, since one can keep the blaster full while the other handles the parts. The best answer to a limited compressed air supply is to rent a commercial-sized air compressor and use this rig while you are doing larger sandblasting jobs like stripping the frame.

Another problem with sandblasting is the large amount of dust it produces. One way to minimize the dust and economize on sand is to build a two- or three-sided pen against which the sandblasting will be done. The pen's walls need only be 3ft or 4ft high and 8ft to 10ft long. Place the parts to be sandblasted in the center of the pen and sandblast toward the walls. Most of the sand and much of the dust will settle inside the pen. The used sand can be scooped up and reused at least once. The residue can be swept up and spread on the garden, lawn, or other spot where it can work into the soil.

Sandblasting is an effective way to get rid of rust on surface areas, but it is not very good at removing rust from inner door cavities, the inside of rocker panels, and other hidden areas. Yet if rust continues to grow in these hidden areas, you're likely to find corrosion blistering through the metal in a few years. Ways to remove or neutralize rust on hidden surfaces are described in the chemical cleaning methods that follow.

The clean, rust-free metal that remains after sandblasting needs to be treated to prevent rerusting—which will occur rapidly if any moisture is present. For short-term rust protection, and to help paint adhesion, freshly sandblasted parts should be washed with metal prep—a dilute phosphoric acid solution available from auto parts stores as a painting preparation product under a variety of names including Metal Prep, Metal Etch, and so forth. Tell the counter clerk that you are looking for a dilute acid product to etch bare metal in preparation for painting, and he or she will know what you want.

You can apply the acid solution with an old paintbrush or rag. Just douse it on liberally—while keeping it off your clothes—and let it run into seams and crevices. When the acid dries, the metal will have a dull gray color.

If you need to store the parts temporarily before painting, place them in as dry a location as possible. If longer storage is planned, the metal should be primed with a nonlacquer-based primer. A lacquer-based primer allows moisture to penetrate to the metal's surface, causing rerusting.

One further caution: Mechanical parts such as the front end or rear end assemblies are not to be sandblasted unless they will be completely disassembled and rebuilt. Never sandblast the engine or transmission because the microscopic dust will coat bearings and other moving parts with a gritty film that will quickly grind the precision metal to scrap.

Chemical Paint Stripping

Chemicals are also used to remove paint and rust. A number of paint-stripping products are available from auto supply stores and specialty restoration supply shops. The paint strippers are either brushed or sprayed onto the old paint layer and allowed to work. When the old paint begins to lift, it is scraped off with a putty knife. The problem

Smaller parts can easily be derusted at home using mild acid solutions. It is important not to immerse suspension assemblies, wheels, or brake drums in an acid solution, since the process can cause a condition called hydrogen embrittlement, which weakens the metal.

is that the paint doesn't scrape clean. Often only one paint layer will lift at a time. This means stripping and scraping each layer. Frequently you'll end up sanding off paint that no amount of stripper seems to loosen.

For factory enamel and urethane paints, you will have the most success using a product called Wet/Dry, which is sold by T. N. Cowan Enterprises in Alvarado, Texas (see the "Suppliers" section of the Appendix). What makes Wet/Dry unique is that it penetrates into the paint and dries, and the paint actually flakes off. Before you conclude that Wet/Dry is all pluses and no minuses, note that this product does not work well with lacquer or finishes that contain various types of paints. It is most successful with factory enamels, which it loosens quite quickly, and eliminates the mess created by other paint-stripping products.

Commercial Metal Cleaning

As an alternative to sandblasting, commercial chemical stripping also removes both paint and rust—as well as Bondo and any other nonmetal coatings. This stripping and derusting process, which places the parts to be cleaned in a caustic solution and uses electrolysis to eliminate rust, is done by commercial establishments that often call themselves "metal laundries." Many belong to the RediStrip franchise chain. To find out if a chemical metal-stripping firm is located near you, look in the Yellow Pages telephone directories of larger cities in your vicinity and in listings that appear in old-car hobby magazines like *Hemmings Motor News* and *Old Cars News & Marketplace*.

The metal laundries clean parts by placing them in large tanks containing caustic chemicals; usually the tanks are big enough to hold items as large as the truck cab or box. The combination of chemical action and electrolysis dissolves paint and eats away rust in anywhere from a few minutes to a few hours, depending on the number of paint layers and the severity of the rust. To protect the metal from rerusting, the commercial chemical cleaners may offer to apply a protective coating of phosphoric acid. Along with retarding rust, this coating will also improve paint adhesion.

Commercial metal cleaning has two advantages over other paint-stripping and rust removal methods. One is that all the surfaces of the part are cleaned, even hidden inside surfaces. The other is that the process does not abrade or eat away good metal. A disadvantage is that you have to transport the parts to

Rust can be removed from sheet metal and large mechanical assemblies using an acid jell. For the acid to work, paint and other coatings must be removed first.

the metal laundry. The major disadvantages of this paint-stripping and rust removal method are that it can be quite expensive—it can cost as much as several hundred dollars to clean the truck cab or box—and the caustic chemicals have been reported to leach out of body seams after painting. If chemicals are trapped in seams where the body panels are joined, they will eventually work to the surface of the metal and lift the paint. If this should happen, the only way to fix the problem would be to try to bake the chemical out of the seam with heat—a process that has no assurance of success.

Acid Derusting

A less expensive chemical derusting process you can do at home uses a mild acid to eat away the rust. Acid derusting works best with smaller parts that can be dipped in a plastic container holding a mild solution of phosphoric acid. Although other acids can also be used to derust ferrous metal parts, phosphoric acid is preferred because it is inexpensive; is less dangerous than other, more caustic acids like sulfuric or hydrochloric; works in a relatively short time frame; and leaves a protective coating that helps paint bond to the metal.

Phosphoric acid in a concentration that is dilute enough to be safe to use is available from local automotive painting supply stores and restoration suppliers like The Eastwood Company, under product names such as Twin Etch and OxiSolv. Instructions on the container give mixing proportions with water—typically 1:1. A lidded plastic pail makes the best container for the acid derusting solution, since the chemical quickly develops a "rotten eggs" smell.

Rusted metal should be completely immersed in the acid solution, if possible, and checked at frequent intervals: every hour or so for a fresh acid bath; older acid works more slowly. Even in its dilute state, the acid will eat away the metal as well as the rust, and if the parts are left in the bath too long, they can be completely destroyed. It's good to try a few scrap parts first, to get used to the speed of the chemical action. Be very careful not to dip nonferrous metal parts (brass or aluminum) in the acid derusting solution because they will be quickly dissolved.

Acid derusting will not work on painted or grease-coated surfaces, so it is necessary first to strip the parts to bare metal. For larger metal parts, such as fenders, that won't fit in an acid derusting bath, acid jells can be used to clean surface rust or even heavier rusting. These jells are available in two forms: common naval jelly, which can be purchased at virtually any hardware store, and Jenolite, which is a somewhat more concentrated acid jell available from T. N. Cowan Enterprises (see the "Suppliers" section of the Appendix).

Like the liquid acid derusting treatment, the jell will not work over paint or grease. Terry Cowan, distributor for Jenolite, recommends first removing surface rust with a dual-action sander. If any corrosion remains, a jell coating can be applied. The acid is wiped onto the metal with a paintbrush. As the jell dries, it leaves a dark gray coating. After it dries, any residue can be wiped off or the part can be "washed" with a dilute metal prep solution. Do not wash the base metal with water. Doing that will almost instantly produce a tell-tale orange coating of oxidation (surface rust). When all rust has been removed, the metal is ready to be primed. (The metal has a uniform gray appearance.)

Although acid solutions and jells provide an effective derusting agent, they also have drawbacks. A major disadvantage is that the acid removes rust by eating into the metal. This means that some good metal is also lost. If the metal is thin, the acid treatment may also destroy the part. Another serious problem is that acid derusting can cause hydrogen embrittlement, a condition that weakens the metal and can cause stressed parts to crack and break. For this reason, suspension assemblies, wheels, brake drums, and other parts on which the car's safety depends should not be derusted using the acid method.

Rust Neutralizing

Rust neutralizing doesn't remove rust, but converts red ferrous oxide (rust) into black magnetite (an inert coating), which protects the metal against further rusting. Most rust converter products, such as

Neutralizers like Zintex convert destructive ferrous oxide to an inert black metal coating called magnetite. One advantage of rust neutralizers is that they can be sprayed into hidden areas like rocker panels and inner door surfaces.

Zintex from T. N. Cowan Enterprises (see the "Suppliers" section of the Appendix), have a watery consistency and can be sprayed, brushed, or poured onto the metal. Areas where the rust conversion treatment works particularly well are inner door surfaces, which can't be reached by sandblasting. Other candidates are chassis parts that you don't want to sandblast and shouldn't derust with acid, wheelwells, the underside of fenders, floor panels, and inside cab panels.

As with acid derusting, it is necessary that the surface be free of grease and paint. If possible, loosen any rust scale with a wire brush. If you are treating the inside surface of the doors with rust converter, it helps to have the doors off the vehicle and placed horizontally—outside-skin-down. Then you can simply pour enough rust converter through the openings in the inner door panel to cover the inside surface. To spread the rust neutralizer over the inside door surfaces and allow it to penetrate into the seams, pick up the door—you may need a helper—and tip it up and down and side to side. After the outer skin is completely coated, flip the door over to coat the inner panel. Some of the rust converter will spill out through openings in the panel, so it's a good idea to spread newspapers on the shop floor in the area where you are doing this, or work out-of-doors.

You can tell that the rust converter has done its job when the rusty metal turns from red to black. Instructions on some rust converter products like Zintex call for washing the metal, then sealing the surface with a coating of paint. Other rust converters like Corroless from The Eastwood Company dry ready-to-paint. For painting inner door surfaces, you can use the same method described above: lay the doors in a horizontal position—outer-side-down; pour primer through the openings in the inner panel; slosh the primer around; let the primer dry; and repeat with a finish coating.

Rust converter penetrates into pits and seams and is ideal where rust removal is impossible or impractical. This process can also be used in combination with other rust removal methods. No one rust removing process is ideally suited for all conditions. More likely, you will use a combination of degreasing, paint stripping, and rust removal to bring your truck to the point where it is ready for

Now that the dirty work is behind you, you can begin to make progress toward your goal: a truck that looks and drives as if it just left the factory.

metal work, refinishing, and mechanical rebuilding.

The rust converter products available from restoration suppliers are safe and nontoxic and do not contain isocyanates, chromates, acids, or lead. These are not experimental products, but have a long history of industrial use.

A Turning Point

Completion of the steps described in this chapter marks the turning point in your truck's restoration or repair. Up to this stage, you have been undoing. From this point on, you will be redoing. Time spent in cleaning and derusting may not be visible in the overall restoration or repair sequence, but these processes are extremely important to the appearance of the finished truck. Of course, cleaned parts are also much easier and more pleasurable to work with. Now that the dirty work is behind, you can begin to make progress toward your goal: a truck you can step out in with pride.

Chapter 6

Straight Axle Front Ends

Starting with the Model T and running through the 1941 models, Ford light trucks—and cars—used a primitive "buggy-style" front suspension that consisted of a solid "beam" axle attached to the front of the frame by a single transverse-mounted semi-elliptical spring. Simple though this suspension was, it gave surprisingly good handling characteristics—with very little lean or roll on corners—and a moderately comfortable ride. In 1942, and through 1965 when Ford introduced its novel Twin I-Beam IFS, Ford trucks used a slightly modified beam axle suspension that replaced the buggy spring arrangement with two semi-elliptical leaf springs.

In either form, the straight axle front end is rugged. This reason probably led Ford to continue to use this suspension design longer than Chevrolet and International. As with any mechanical system, however, a straight axle front end is also subject to wear, and on most older trucks, both the front suspension assembly and the steering linkage are likely worn enough to require a complete rebuild.

The steps that follow describe the basic procedure for overhauling a pre-1965 Ford light-truck front end. Specifics may differ, so you are also

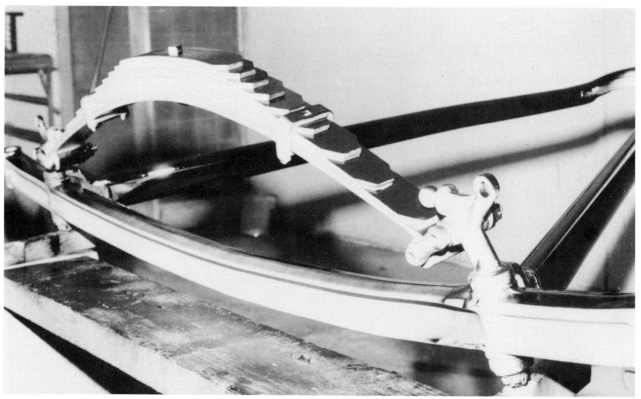

Ford cars and light trucks through 1941 used a buggy-style front suspension that consisted of a solid beam axle attached to the front of the frame by a single, transverse-mounted spring. One reason Ford kept this primitive suspension system for so long was that it gave surprisingly good handling and a moderately comfortable ride.

advised to refer to a service manual for your year and model truck. The intent of these instructions is to give you an overview of the rebuilding sequence and to point out some of the difficulties you are likely to encounter in working on parts that may be nearly as old as, or older than you.

Common Causes of Problems

A straight axle front end gets its name from the axle beam that connects both wheels and supports the spring or springs that cushion the driver and cargo from bumps and road shock. The springs also hold the axle in place. A beam axle is little more than a forged piece of quality steel with holes drilled in the ends. Pins fitted through these holes allow wheel assemblies to pivot as controlled by the steering mechanism.

The many moving parts of a straight front axle assembly require periodic lubrication, which on most older pickups, they didn't get. The resultant wear shows up in sloppy steering that may become loose enough for the front wheels to chatter and shimmy after hitting a bump and for the truck to wander down the road rather than hold a straight course.

Sloppy steering has a number of causes: worn kingpins or bushings, worn ends on the tie rod (the long rod that connects the two wheels), loose wheel bearings, wear or incorrect adjustment in the steering box, or a combination of these conditions. Along with loose steering, other front end problems include a slouching front stance (the truck tips to one side or seems to settle lower than normal in the front) and rapid tire wear. A slouching front end traces to sagging or broken springs or worn spring shackles. Rapid tire wear can be caused by a bent axle or incorrect tie rod adjustment that pitches the tires toward each other (a condition called toe-in) or away from each other (a condition called toe-out).

Skills and Tools

Rebuilding a straight axle front end is within a weekend mechanic's skills but requires a rather complete mechanic's tool set, plus a few specialty items, and a shop manual, which will give an overview of the various repair sequences and is needed for toe-in and other settings. At a minimum, the tool set should consist of the following:
1/2in socket set
hefty ball-peen hammer
assorted chisels
pliers
screwdrivers
punch set
bushing drivers
bushing reamer (optional)
gas torch (optional, recommended)
puller, tie rod separator (optional, recommended)
bench vise

Assessment and Diagnosis

Overhauling the front end should begin with checking its overall condition and diagnosing its problems. To do this, you will perform three simple tests. For the first test, grip the steering wheel and turn it back and forth slightly. As you do so, look for movement of the front wheels. If the steering wheel has a noticeable amount of free movement—called "play"—through which the front wheels are stationary, the tie rod ends and steering linkage may be worn, the steering box may need adjusting or may have internal wear or damage, or a combination of these conditions may exist.

The second test requires that you park the truck on a level surface—which in most cases can be the garage floor. Here you will check for sagging or broken front springs and worn shackle bushings. This is done both visually and by measurement.

For the visual check, sight along the springs. Are the springs curved in an arc, or are they essentially flat? Most front springs had some arc when new. Now closely inspect the individual leaves for breaks. Also look for misalignment, which is another clue of a broken leaf. Often broken leaves can't be spotted until a spring is disassembled, but if one spring is flatter than the other, or has more sag, you'll

From 1942 through 1965, Ford trucks used a slightly modified beam axle front suspension that replaced the "buggy spring" arrangement with two semi-elliptical leaf springs.

55

probably decide to remove and overhaul both springs. If you are taking the rebuilding approach and the springs seem to have more or less normal arc and don't show any broken leaves, you will probably leave well enough alone. On a frame-up restoration, you will remove and disassemble both front springs; have the leaves rearced; then paint, lubricate, and reassemble the spring assemblies. While visually inspecting the front suspension, also look for worn spring shackle bushings. The spring shackles are the brackets that attach the ends of the springs to the frame. Wear can usually be detected by space for the shackle bolts to move inside the bushings.

As a sure test for spring sag and worn shackles, measure the distance from the floor to each frame horn; for these measurements to have meaning, it is very important that the floor be level. The difference in the two measurements is the amount of spring sag and possible shackle wear.

The third test checks for looseness anywhere in the front end assembly. To do this, jack the front end off the ground and place a jack stand under either end of the front axle. Now lower the truck onto the jack stands, make sure the stands are supporting the truck's weight, and block the rear wheels to keep the truck from rolling. You can begin checking for looseness by inspecting the U-bolts that hold the springs to the axle. Sometimes a U-bolt is broken or loose. A broken U-bolt can allow the front axle to move as the truck is going down the road. Also inspect the bolts holding the steering box to the frame. If the steering box is loose, you can expect to feel play in the steering.

Next check the wheel bearings and kingpins for looseness by gripping the tire with one hand on the top and the other on the bottom. Now attempt to "rock" the tire by pulling one hand toward you while pushing the other away. More than the smallest amount of free movement—about 3/16in is tolerable—indicates that either the wheel bearings or kingpins, or both, need service.

Finally grab hold of one end of the tie rod and forcefully pull the rod up and down to check for play. Now do this at the other end. If either end moves up and down, the connectors are worn and will need to be serviced or replaced. If you twist the tie rod, you may feel some rotating movement; this is normal. Make the same checks for looseness on the ends of the drag link—sometimes also called the steering connecting rod.

After performing these checks, write down all the places where you have noticed looseness. Front end assemblies on older trucks typically show extensive wear owing to infrequent lubrication and hard use over rough, unpaved roads. If your goal is restoration, you will probably decide to overhaul the front end completely. If your goal is to make the truck serviceable, you now know the areas that need attention. The next sections describe how to bring a straight axle front end back to safe operational standards.

Front End Disassembly

Typically the front end overhaul begins at the point in a frame-up restoration where the front end sheet metal, cab, and box have been removed, exposing the bare chassis. The engine and drivetrain have also been pulled, leaving just the frame and front axle assembly. To remove the axle and its attached suspension and steering components from the frame, you will start by disconnecting the shock absorber link on lever-style shocks, or the nut from the lower rod on double-acting shocks. If the truck is equipped with a stabilizer bar, remove this next. Now you can disconnect the brake flex hoses—on trucks with hydraulic brakes—from the front wheel cylinders. In most cases, these hoses will have dried out and cracked and will need to be replaced later during brake overhaul. If the hoses are in this condition, the simplest way to get them out of the way is just to cut them. On an earlier truck with a mechanical brake system, you will disconnect the front brake rods or cables.

Now the drag link needs to be removed. This is the short rod on the left side that attaches to the pitman arm—also called the steering sector shaft arm—from the steering box. The drag link attaches to the pitman arm at the steering box and to the left spindle arm. On Ford cars and light trucks from 1935 and later, the drag link is removed by prying the tapered end bolts loose from their sockets in the pitman and spindle arms using a tie rod separator. On Model Ts and As and early V-8 Fords through 1934, the drag link is removed by pulling out the cotter keys, loosening the plugs, and then prying out

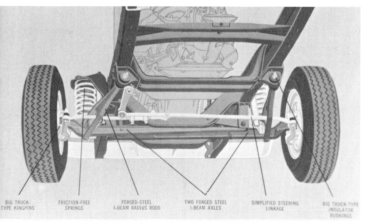

In 1965, Ford introduced a modification of the beam front axle, that allowed each front wheel to respond to bumps and irregularities in the road surface independently of the other. Ford called this suspension Twin I-Beam, and indeed it did contain many features of the earlier beam axle designs.

the ends of the steering sector shaft and spindle arms. The ends of drag links that have this plug and seat design are repairable, and repair kits are readily available from vintage Ford truck parts suppliers.

The final step in removing the front end assembly from the frame is to detach the springs by taking apart the spring shackles. If the shackle bolts are rusted, they can be heated or cut with a torch. You will be replacing the spring shackles, so any damage to these parts is of no consequence. With the spring shackles apart, the front axle assembly can be moved away from the frame to an area of the shop where it can be worked on further.

Front Axle Disassembly

In the next series of steps, you will break the front end assembly down into its individual parts. When this is done, you can clean and degrease the parts, sandblast as needed, and then begin the rebuilding and refinishing process.

Spring Removal

You will start the disassembly by removing the springs (it is assumed that the wheels have been removed and that the frame is resting on supports). This is done by loosening the U-bolts that hold the springs to the axle or, on pre-1942 Ford light trucks with the single front spring, the spring to the frame. Typically the nuts to these bolts will be rust frozen. Squirt plenty of penetrating oil on the nuts before attempting to loosen them. If you have a torch, heating the nuts will ensure they come off easily. Whenever possible, avoid damaging or destroying the original nuts and bolts because modern replacements rarely match the originals in bolt head size and appearance.

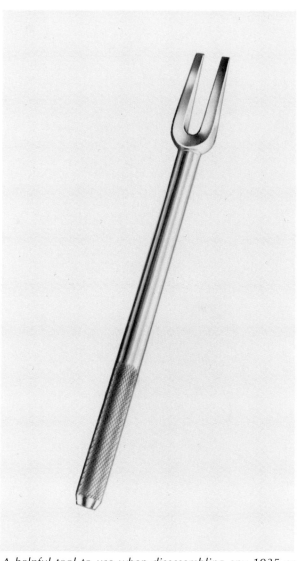

A helpful tool to use when disassembling any 1935 or newer Ford front end is this tie rod separator.

The first step in removing a beam front axle is to disconnect the shock absorber, followed by the brake flex hoses. In most cases, the brake flex hoses have dried out and cracked and will be replaced later. When this is so, it's easiest just to cut the hoses.

The easiest place to disassemble the front suspension is out from under the truck. Air wrenches make fast work of removing nuts and bolts. A hefty hammer may be needed to drive out the spring shackles and kingpins. Chrystal Edgerly

Before the kingpins are tackled, the locking pins that fit through holes near the ends of the axle have to be removed.

When the U-bolts have been removed, you can further disassemble the springs by loosening the center bolt. You also need to remove the bushings from the spring eyes. This is best done with a bushing driver. Keep the spring sets separate by tying together the leaves from each spring or placing each set of leaves in a separate box. Label the springs Left or Right side, as appropriate. Later, when the front axle system is completely disassembled, you will clean and sandblast the spring leaves and have them rearced. For now, you just want to make sure they don't get misplaced or mixed up with other parts.

Tie Rod Removal

Now you can remove the tie rod (the long rod running parallel to and just behind the front axle). To loosen the tie rod, remove the cotter keys that fit through the hex nuts on the studs that attach the tie rod to the spindle arms. Since the sockets on the spindle arms are tapered, some force is usually required to free the tie rod end bolt from the socket. At least three methods can be employed to pop the bolt loose. The slowest is to use a puller. The fastest is to drive a special two-pronged tool called a tie rod separator between the tie rod end and the spindle

Often the kingpins are rusted to the axle bores and need some "persuasion" to be driven out.

To remove the kingpins with the front axle still in the truck, I have sometimes found it necessary to use a portable hydraulic press. Even with the 5 tons or more of pressure exerted by the press, it has occasionally still been necessary to heat the axle ends before the kingpins would pop loose.

arm. This tool is available from most auto parts stores. The third method, which is used when you don't have either of these tools, is to wedge a pry bar between the tie rod and spindle arm, press down on the bar to put some pressure on the tie rod bolt, and rap the bolt with a couple of sharp blows from a ball-peen hammer. If you use the pry bar method, loosen but don't remove the end of the bolt. If you pound directly on the threads, you may spread the end of the bolt so that it won't fit through the hole in the spindle arm. When you have done this procedure at both tie rod ends, the rod can be pulled free.

Brake Removal

To get at the wheel spindles, which attach to the yokes at the axle ends, you first have to take off the brake drums, disassemble the brakes, and remove the backing plates. The brake drums should slide off rather easily after you loosen the spindle nut and remove the outer wheel bearing. Sometimes, however, the brake shoe contact area has worn so that a lip is formed on the edge of the drum, and this lip catches against the brake linings. If this is the case, you will have to loosen the brakes—back off the shoes—before the drum will pull loose. (For instructions on this procedure, refer to chapter 9 on brake overhaul.) Be sure to use a wrench, not pliers, to turn the spindle nut. The serrated jaws on a set of pliers will cut into the nut, making it difficult to fit on a wrench when you or someone else decides to use the right tool. Also be sure to keep the bearings and all parts for each side together.

With the brake drum removed, it's a good time to pull the inner wheel bearing out of the drum. This is best done by using a special bearing puller. You can also remove the bearing by prying the bearing seal out of its groove and driving the bearing out of the race by gently tapping on the bearing cage with a punch. Another, easier method for removing the inner wheel bearing is to replace the nut and washer on the spindle, then slip the brake drum over the nut and let the hub rest on the spindle. Now pull the drum toward you with a swift, sharp tug striking the inner bearing against the spindle nut. This will pop the bearing free without damage to the cage, nearly every time.

Before removing the brake shoes and related parts, it's a good idea to take photos of the brake assembly. These photos may be useful in

59

reassembling the brakes later.

To disassemble the brakes, begin by removing the springs and clips holding the brake shoes in place. You should be able to work through this mechanical puzzle fairly easily. Be sure to save all parts.

With the brake shoes out of the way, you will see the bolts holding the backing plates to the wheel spindles. Remove the cotter keys from the bolts at the nut end, and loosen the nuts. The backing plates will then slide off the spindles. Set the bolts and nuts aside for plating.

At this point, you are ready to drive out the kingpins—also called spindle bolts—and remove the wheel spindles from the axle.

Kingpin Removal

The kingpins fit through holes in the spindle yoke and axle. They are held in place by wedge-shaped locking pins that are driven through a small hole near the end of the axle and kept from working loose by a nut that is usually found on the back side of the axle. To remove these locking pins, loosen the nuts and drive the pins out of their holes with a punch.

You may also find dust seals at the top and bottom of both kingpins. These are metal disks that have been wedged into the holes in the spindle yoke. It is not uncommon for these seals to be missing, either because they were not installed during an earlier overhaul or because they have worked loose and fallen out. The dust seals are made of soft metal and can be removed easily by forcing a small, sharp chisel into them and prying them out of the hole.

With the seals removed, you can attack the kingpins. Sometimes kingpins will slide out of the axle as easily as a knife slips through soft butter, but more often, they are rusted to the axle and move about as willingly as a knife saws through frozen butter. Before attempting to drive out the kingpins, spray penetrating oil into the openings at the bottom and top of the spindle yoke, the locking pin hole, and the openings between the axle and spindle. It's difficult to get the penetrating oil directly into the hole in the axle end where the kingpin is likely to be rust frozen, but try.

After letting the penetrating oil work, place the end of the axle on a cement block, anvil, or other stationary object that will not absorb the force directed against the kingpin, and attempt to drive the pin out of the axle using a hefty 24oz hammer and a punch that is slightly smaller in diameter than the pins.

If the pin moves after several sharp blows, apply more penetrating oil; fortune is smiling, and by alternating hammer blows on the pin with squirts of penetrating oil into the openings, you will soon have the kingpins out and the axle assembly completely apart. More likely, however, the kingpin won't move in the least. If this happens, you have two alternatives. Assuming that the front axle is out of the truck—as has been described in the steps above but may not be the case if you are using a repair approach—you can take the axle to a machine shop and have the pins pressed out on a hydraulic press. The other alternative, which you will have to use if the axle is still in the truck, is to heat the axle ends with a torch until the metal glows red. A propane torch won't apply enough heat; you will have to use an oxyacetylene or MAPP gas welding torch. Once the metal around the pins has been heated, you should be able to drive the pins out using strong, sharp blows, as described earlier. You won't be reusing the kingpins from your truck, so you can throw them out, along with their related hardware.

With the front axle disassembled, you will want to clean it for repainting. Since sandblasting won't remove grease and caked-on mud, these coatings will have to be scraped off using the manual method shown here, or cleaned off in a degreasing bath. Chrystal Edgerly

One of the tools needed to replace the kingpins is a bushing reamer. This is used to match the inside diameter of the new spindle bushings to that of the kingpins.

Front Axle Parts Cleaning, Inspection, and Preparation

Now the axle assembly should be completely apart. Either at this point or after you have cleaned the spindles, you will need to remove the kingpin bushings from the spindle yokes. This is best done with a bushing driver. An alkaline degreasing bath (described in chapter 5) will make short work of the accumulated grease and dirt commonly found on the front axle and its associated parts. Rust, which is likely to be found on the axle and brake drums, can be removed by sandblasting or treated with a rust neutralizer. Front end parts should not be derusted using acid because of the danger of embrittling the metal.

After the axle has been cleaned, check it for trueness. With the rough use most trucks have experienced, it is not unusual to find that the front axle has been twisted or bent. You can test for twisting by laying the axle on its side on the shop floor and looking to see if both axle ends touch or are the same distance from the floor (this assumes that the floor is smooth and level). You can check for bends by sighting along the axle. If either twisting or bends are noticed, take the axle to a machine shop and have it straightened.

The tie rod should also be checked for straightness. If this long rod is bent or kinked, you may want to consider finding another that is straight and true. The other option is to straighten the rod you have. This should be done "cold." Heating the rod to straighten it will cause the metal to lose its temper and allow it to bend more easily in the future.

While inspecting the tie rod on Ford trucks through 1934, also check the condition of the spindle arm balls that fit into the sockets on the ends of the tie rod. Few trucks are maintained as conscientiously as cars, and if lubrication has been neglected, the spindle arm balls have probably worn into an egg shape. Out-of-round balls will cause the truck to be hard to steer and can also contribute to play in the steering. If your truck's spindle arm balls are no longer round, you have the choice of replacing the spindle arms or having new balls welded onto the arms. These balls are available from vintage Ford parts suppliers. You should have the work of cutting off the old balls and welding on the new ones done at a machine shop. Strong welds are critical. If a weld should fail and a spindle ball break off on the road, the truck would go out of control.

If you haven't taken the springs apart, this needs to be done next. Removing the center bolt is all that's necessary to separate the leaves. The leaves can then be cleaned and derusted—either by sandblasting or by wire brushing. Before reassembly, the springs should be rearced. This is done by a spring shop. After rearcing, the leaves can be primed and painted. Before reassembling the springs, place either strips of Teflon or a liberal coating of grease between the leaves. This is done to enable the leaves to slide freely on one another, and greatly improves the ride. After the center bolt has been replaced and tightened and new bushings have been installed in the eyes at the ends of each main leaf, the springs are ready to be installed on the axle.

All front end parts should be primed and painted in preparation for reassembly. While refinishing is going on, you can send the hardware (nuts and bolts) out for zinc plating. Any commercial plating shop should be able to provide this service. At this time, you should also draw up a parts list and begin gathering the items you will need to reassemble the front end assembly. These will include a kingpin set; tie rod ends; drag link end assembly kits—and replacement spindle arm balls where needed; new or relined brake shoes; brake return springs; wheel cylinders; spring shackles; and wheel bearings, which should be replaced if the old bearings show signs of wear or spackling (indicated by chips in the rollers) or if the inner wheel bearings have broken apart in the process of removing them from the brake drums. You will also need new inner wheel bearing seals.

Although you can get all of these items from the vintage Ford parts suppliers listed in the Appendix, you may also be able to purchase some of them at a nearby NAPA auto parts store. Buying parts locally has the advantage of saving shipping costs and delays, and if the items aren't right, you can easily return them. Owing to NAPA's enormous warehouse inventory, parts listed but not in stock can usually be shipped to a local store within one or two days.

Front Axle Reassembly

Once the chassis and other front end parts have been refinished, you will set about rebuilding the front end in the reverse order of its disassembly. This means that the new kingpins will be installed first.

Kingpin Reinstallation

You have basically three things to remember when installing kingpins:

1. Make sure the holes in the bushings are in line with the holes in the spindles for the grease fittings.

2. Place the thrust bearing in the correct location—between the axle and the lower spindle yoke.

3. Shim the space between the axle end and the spindle yokes so that there is no more than 0.005in of gap and no free movement.

If you follow these guidelines, you will have accomplished the first step in ensuring that your truck will steer easily and run down the road straight and true.

The procedure for replacing the kingpins begins by installing new bushings in the spindle yokes. This is best done with a bushing driver, available from a specialty supplier like The Eastwood Company, and an arbor or hydraulic press. Most hobbyists don't have a press, and if that's your case, you may choose to have these bushings installed at a machine shop. If you'd rather put them in yourself—because no machine shop is nearby or because you like to do your own work or to save money—you can press the bushings in place using a bench vise.

When using a vise to press the bushings, first you need to start the bushings in the holes in the spindle yoke. To start the bushings, round their edges slightly with a file, squirt some oil into the hole in the spindle yoke, and then tap the bushing into the hole with a plastic hammer. It's a good idea to place a large washer over the end of the bushing you are tapping to make sure no damage is done to the bushing's soft metal. Be sure the hole in the bushing lines up with the grease fitting hole in the spindle yoke. Once the bushing is started, you can place the spindle yoke in the bench vise—you'll need to insert a block of wood between the bushing and the vise jaw to prevent the jaw from cutting into the soft metal of the bushing—and slowly press the bushing into its hole by tightening the vise.

When all four bushings have been installed, (remember to make sure the bushings are aligned with the grease fitting holes) they need to be reamed to the diameter of the kingpins. You can do this yourself if you have a reamer of the correct diameter, which can be purchased from an auto parts store or specialty tool supplier, or you can have the bearings reamed at a machine shop. The cost of having a machine shop do this work will be about the same as the cost of the tool, so whether you decide to do the work yourself or hire it out will probably depend on whether or not you want to add a bushing reamer to your tool inventory. If you choose to buy the tool, be sure to get a reamer that is long enough to reach both bushings. This is necessary to establish proper alignment for the kingpins.

To assemble the kingpins, first fit a spindle yoke

The drag link and tie rod ends on Ford cars and light trucks though 1934 contain a socket assembly consisting of a spring set, a spring, two ball seats, and an end plug.

When the parts are assembled, you keep the end plug from loosening by inserting a cotter pin through holes that align with a slot in the plug.

over one end of the axle. Next place a thrust bearing between the axle and the lower yoke. Be sure that the closed side of the thrust bearing is at the top. Now check the clearance between the upper face of the axle end and the upper yoke. This gap should not exceed 0.005in. If a greater distance exists, insert shims in this space until the required clearance is met. As you install the shims, check the gap with a feeler gauge. Too loose a fit will cause sloppy steering, but too tight a fit will make for hard steering. As you fit the shims, be sure the holes in the spindle yokes line up with the hole in the axle. Once the shims are in place, you can fit the kingpin into the hole. Make sure the notch for the locking pin faces toward the locking pin hole in the axle. With the kingpin in place, drive the locking pin through its hole in the axle until it wedges tightly against the kingpin. Now install a lock washer and nut on the locking pin's threaded end and tighten. To seal the kingpin against dust and moisture, place a soft metal plug, rounded-side-up, in the hole above the kingpin on one end of the spindle yoke and wedge the plug in place by giving it a sharp tap on the crest of the bulge with a ball-peen hammer. Do the same at the other spindle end. Now repeat this procedure with the other spindle, and the job of installing the kingpins will be completed.

Tie Rod Reinstallation

If the tie rod ends are worn, the new ends need to be threaded into the tie rod. Be sure the threads are cleaned and well lubricated, as it is important that the rod be able to turn easily in order to set the front wheel toe-in. Before threading on the ends, slide the clamps that lock the ends in place onto the tie rods. Do not turn the ends fully into the tie rod, but leave about one-third of the threads showing.

Ford light trucks since 1935 use replaceable tie rod ends that thread into the tie rod and fit into a tapered hole in the spindle arms.

Position the ends so that roughly the same number of threads are showing in front of each.

On the front end of Ford light trucks from the mid-thirties, the tie rod ends contained a socket assembly consisting of a spring seat, a spring, two ball seats, and an end plug. Assemble these parts in the sequence shown in the service manual, with the spindle arm ball sandwiched between the seats. After the parts are assembled, tighten the end plug until the spring is fully compressed, then back off to fit the cotter pin through its hole. Lock the pin in place by bending back the ends. Now repeat the same procedure at the other tie rod end.

Spring Reinstallation

Now you are ready to mount the front springs. On later models, this is done by centering the springs over the axle and clamping them in place with U-bolts. On the earlier transverse spring front ends, the spring attaches to the axle by the spring shackles; you will need a spring spreader to stretch the spring in order to fit the shackles.

Front End Reassembly

If the frame has been cleaned, primed, and painted, you can now reattach the front axle assembly. The transverse spring front ends attach by replacing the U-bolts; on the later front ends, the springs are attached by replacing the spring shackles.

If the steering box is also in place, you will also reattach the drag link. On Ford light trucks from 1935 and later, the drag link had one replaceable and one fixed end. If the fixed end is worn, a new drag link will be needed. On pre-1935 Ford trucks, remember to inspect the drag link's internal connecting parts (ball seat, springs, and so forth) and replace them if necessary. The shock arms also need to be connected to their mounts on the frame.

Steering Adjustment and Overhaul

Although rebuilding the front end usually removes most of the steering play, wear in the steering box may also be contributing to front end looseness. Normal steering box wear can be taken up in adjustment, but if the truck has been driven with the steering box low on lubricant or is a high-mileage vehicle, the steering box may need to be rebuilt or replaced. Looseness that appears to be coming from the steering box can also be caused by loose bolts holding the steering assembly to the frame. Check the steering box mounting bolts for tightness.

Play in the steering box can be detected by having an assistant hold the pitman arm at the bottom of the steering box while you turn the steering wheel. If free movement is noticed while the pitman arm is held rigid, you may be able to take up the slack by adjustments. If not, the steering box will need to be rebuilt or replaced by a tighter unit.

In some cases, the steering box may be free of

play but excessively high effort may be required to turn the wheels. Causes here may be little or no lubricant in the steering box; improper steering adjustment; or overtightened tie rod ends, on trucks with adjustable-type tie rod sockets. Another cause of hard steering is underinflated tires.

Adjusting steering play begins by disconnecting the steering sector shaft arm from the pitman arm. It is important to note the relative position of these two parts before disconnecting them so as to get them back in the same position.

The rest of the procedure will be done at the steering box. Two adjustments will be made. One corrects for slop that has developed between the worm gear and the roller bearings that support the worm gear and its attached steering shaft. The second adjustment reestablishes proper alignment between the worm gear and the sector (the shaft and sawtooth-shaped gear that turn the pitman arm and steering linkage). When both of these adjustments are set correctly, the steering should have the correct pull (between 1/4lb and 3/4lb) and no play should be felt in the steering box. Of course, it is possible for these adjustments to be set correctly and the steering still to be excessively hard and play still to exist. With a truck that has seen hard service and little maintenance, likely causes of steering box problems iNclude worn sector bushings, worn steering shaft bearings, worn worm and sector gears, and a bent sector shaft. Since Ford steering boxes varied over the years, you should refer to the service manual for your truck when making the steering box adjustments.

After adjusting the steering box, reattach the steering connecting rod to the pitman arm and again check for play in the steering. This is best done by jacking the front end off the ground, and having a helper hold a front wheel while you turn the steering wheel. If the tie rod ends and kingpins are tight, which they should be if the front end has just been rebuilt, any free movement in the steering wheel that is not felt at the front wheel means that the steering box needs to be either rebuilt or replaced.

Instructions for removing and rebuilding the steering box are found in the service manual for your year and model Ford truck. Rebuilding the steering box is not especially difficult. Unless you are disassembling your truck for restoration, however, you will find that removing the steering box and column assembly is a rather complicated procedure that also requires pulling the steering wheel. For this job, you will need the correct tool: a steering wheel puller, which may be available at a local tool rental. Under no circumstances should you attempt to remove the steering wheel by pounding against its base. Doing so will crack the plastic and quite possibly destroy the wheel.

If adjusting the steering box appears to have

Steering boxes on vintage Ford trucks may not be as simple as the Model A steering box disassembled here, but they are not overly complex. If the steering box on your truck shows excessive wear after all adjustments have been made, you may decide to rebuild it yourself following the step-by-step instructions from a Ford service manual.

eliminated excess play, be sure the box is filled with lubricant before operating the truck on the highway. Ford owner's manuals call for filling the steering box with Society of Automotive Engineers (SAE) 90-weight gear oil. The lube is pumped into the steering box through the filler opening located in the upper side of the box, which is sealed with a pipe plug. Do not install a grease fitting in this hole. Grease should not be put into the steering box under pressure. When these trucks were in regular use, mechanics often added a small amount of chassis lube to the 90-weight gear oil in the steering box largely to help prevent the fluid oil from seeping past the seals. Although the manuals do not call for doing this, you may want to fill the box nearly full with gear oil and lube and top it off with chassis lube.

Setting Toe-in

With the front end rebuilt and the steering box adjusted or rebuilt so that all play is out of the steering, the last step remaining is to set the front wheel toe-in.

For directional stability, it is necessary that the front wheels point toward each other slightly in the front. The degree to which the front wheels are angled inward is called toe-in. If the wheels point out, the condition is called toe-out and results in very rapid tire wear. At least two methods, not requiring special tools, can be used to set the front wheels of your truck to the correct "toe."

With the first method, you need to jack up the truck so that the front wheels are slightly off the shop floor. Place jack stands under the front axle, and be sure to block the rear wheels so that the truck will not roll. Now spin a front wheel with your hand and spray a thin line of paint—use a light-colored aerosol spray—along the center of the tire tread. Do the same with the other front tire. When the paint has dried, spin each wheel again and scribe a line in the center of the tire using a nail, Phillips screwdriver, or similar pointed tool. Next lower the truck so that the front tires rest on the floor. Now measure to the midpoint of each front tire and mark an X across the line. Make this mark at the same height on the front of both tires, and then do the same on the back of both front tires. Using a carpenter's tape, measure the distance between the Xs at the front and at the back of the tires. The measurement from X to X across the back should be slightly greater than that across the front.

Typically the correct toe-in has a measurement at the back of the tires between 1/16in and 1/8in greater than that at the front. You will, however, want to check the service manual for your truck for the exact toe-in specification. If your measurement is greater or less than the specified toe-in amount, you will need to jack up the truck slightly and adjust the toe by loosening the tie rod ends and turning the tie rod. To check the toe-in, you will once again need to lower the truck and take another set of measurements. It may be necessary to repeat this process several times to get the correct toe-in setting.

The second home-shop method for setting toe-in uses a gauge comprising two sticks of wood clamped together (1in stock works well for this). If this two-stick wooden gauge sounds too crude and simple, you can buy an inexpensive-yet-precision toe-in gauge from The Eastwood Company. With the front tires resting on the shop floor, loosen the clamps and extend the sticks, or the metal rods of the toe-in tool, until they touch the outermost tread about halfway up the back of the tires. Now tighten the clamps so that the gauge is set to this length; using the toe-in tool, this measurement is made to the outer edge of the tire. Next move the gauge to the front of the tires and check the amount of toe-in. As noted in the previous procedure, the correct toe-in is typically a difference of 1/16in to 1/8in greater distance at the

Wear in a Twin I-Beam front end is detected by jacking up the front end and moving the radius arms and axle beams with a pry bar to check for looseness in the bushings at the axle pivot points and frame connection.

back of the tires than at the front.

After the toe-in has been set, torque the bolts on the tie rod end clamps to 100 pounds-feet (lb-ft) to 120lb-ft. Be sure to install cotter pins to ensure that the bolts don't work loose.

1965 and Later Twin I-Beam IFS Rebuild

In 1965, Ford made a major advance in light-truck suspension by developing the Twin I-Beam IFS. Ford cars had acquired IFS in 1949, but Ford trucks had retained straight axles owing to their simpler, more rugged design. IFS gives a ride much superior to that of a straight axle, however—which inspired the expression "rides like a truck"—so Ford engineers faced the challenge of how to build a suspension that would take hard use without coming out of alignment and causing premature front tire wear. The resulting Twin I-Beam design has proved so rugged that it is still used today.

Twin I-Beam IFS is an ingenious modification of a straight axle front end. As the name implies, Ford's engineers achieved IFS by using two straight axles, which allow each wheel to ride over bumps independently of the other. Rather than leaf springs, the Twin I-Beam suspension uses coil springs. Other features of the earlier straight axle front end, including kingpins, are retained.

Wear points on the Twin I-Beam suspension, besides kingpins and tie rod ends, are the bushings at the axle pivot points and frame connection. With many miles and the hard use trucks are commonly subjected to, these bushings do wear out. Bushing wear is detected by jacking up the front end—be sure to support the truck on jack stands—or raising the truck on a hoist and moving the radius arms (the stubby arms that attach to the frame) and axle beams with a pry bar. Movement is checked at the bushing end of the radius arm or axle. If excessive radius arm or axle movement is noted, the bushings are worn and new ones need to be installed. Instructions for replacing the bushings are shown in the Ford truck shop manual. Rebuilding a Twin I-Beam front end is not a lot more complicated than rebuilding a beam axle, but requires special tools for compressing the coil springs and replacing the bushings.

Although Ford's Twin I-Beam front suspension is a rugged design, bushing wear is its weak point. If your Twin I-Beam truck shows rapid front tire wear, the bushings are likely candidates for replacement. If the bushings appear tight, adjustments may be needed in toe-in or camber. These should be set at a front end shop.

Tip to Improve Steering Ease

The steering linkage of an older Ford pickup has a lot of friction points. Drilling and tapping each end of the drag link for extra grease fittings so that each side of the steering ball gets grease can reduce friction and lighten the steering effort.

Chapter 7

Engine Overhaul

Restoring an old truck is more than building a vehicle that looks nice; you also want it to run like a well-oiled watch. Given the mileage many older trucks have traveled and the hard work they've seen, an engine overhaul is likely to be in order—and will almost certainly be done if you are restoring your truck for show competition. Engine accessories—the carburetor, generator, and starter—are likely candidates for overhaul as well.

Skills and Tools

The question you may be asking yourself about this mechanical work is whether you feel capable of doing the overhauls yourself. If you're adventuresome enough to tackle rebuilding your truck's steering and front end, overhauling a carburetor or the generator and starter will easily be within your mechanical talents. Rebuilding an engine requires extreme attention to detail and knowing how to assess the condition of the internal parts, as well as a basic understanding of how to put everything back together. This isn't to say that overhauling an engine is beyond the skills of a hobbyist-restorer; it can be done, but you're likely to need some help—or at least advice—from a seasoned mechanic along the way.

The purpose of this chapter isn't to take you through the engine overhaul process step by step; you'll find that kind of instruction for rebuilding and reassembling an engine in the Ford shop manual for your truck. Rather this chapter describes how to check the engine's condition and gives guidelines on having required machining done. This is the information you'll need to build your confidence and get started.

Mechanical repair work is much less daunting if you can draw on the experience and expertise of a seasoned mechanic. You can choose from several ways to plug into an expert's help. The easiest and most enjoyable is to join a club where you can meet others who have an interest in older vehicles. In any group of this sort, you will usually find enthusiasts with strong mechanical skills who will take an interest in your project and be willing to lend advice, if not help.

Another way to find expert help is to sign up for an auto mechanics course taught at a local vocational education center. You won't become a master mechanic in ten to fifteen weeks, but the course will give you a good foundation for the mechanical work needing to be done on your truck and you will be able to draw upon the instructor's expertise for answers to questions you may have with your restoration project.

Still another way to get help when you find yourself over your head on some mechanical project is to develop a friendship with a retired mechanic. Such a relationship can benefit you both.

The experience of overhauling your truck's mechanical assemblies can be richly satisfying. You will need a mechanic's tool set (described in chapter 4) plus some specialty items such as pullers, an engine stand, and a torque wrench. You will also need a collection of shop manuals. Besides the Ford shop manual for your truck, it also helps to have the *Chilton's* or *Motor* manual that covers your year

If you are tackling major mechanical work for the first time, you will feel much less pressure if you can draw on the assistance of a seasoned mechanic.

truck. The reason for needing more than one service manual is to get different views of the disassembly, repair, and reassembly steps. The various manuals also describe the overhaul sequences in their own way, giving you a better understanding of what you are trying to do. Instructional manuals—used in automotive courses—on rebuilding engines, transmissions, rear ends, electrical accessories, and fuel system components are also helpful. You can sometimes find these at used-book stores or swap meets. With these books, you'll learn the theory that is missing from a service manual.

If you're not yet confident of your mechanical skills, the place to begin is with the engine accessories. The carburetor, generator, and starter are mechanically quite simple and can easily be overhauled inside the house as cold weather projects. Unless your truck has exceptionally low mileage or these components have recently been rebuilt, some maintenance or repair is most likely in order. As you follow the guidelines below for rebuilding these engine accessories—along with referring to the step-by-step overhaul sequences in your truck's shop manual as needed—you will soon find your mechanical skills and confidence growing to the point where you will be ready for the challenge of bigger mechanical projects—like overhauling the engine.

Carburetor Overhaul

The single- or dual-throat carburetors, used on Ford flathead V-8s and six-cylinder trucks, are the essence of simplicity. But, as is common with low-tech designs, these simple fuel-metering devices can appear to operate just fine while actually needing a rebuild. Usually engine performance deteriorates slowly; the points wear, the ignition timing slips, the carburetor begins to operate a little rich, the engine may begin to start a little hard, and fuel economy drops off. Those of us who have found ourselves chasing the clock for several weeks running know the feeling. We keep cranking out the work that's expected of us, then one day, we punch our "go" button and nothing happens. If we're smart, we take a day or two off and get recharged. An engine running in less-than-top condition doesn't ask for time off, but giving it a little attention to get its systems back in repair is a good idea.

Whether you are concerned about just making your truck run better or doing a complete restoration, rebuilding the carburetor is part of the process. Although this chapter concentrates on the simple one-barrel carburetor layout, rebuilding a two- or four-barrel carburetor follows the same principles. This section presents an overview of the rebuilding process and an insight into how a carburetor works. When you are doing the actual rebuilding, you'll want to follow the step-by-step instructions in the Ford shop manual for your truck.

Carburetor Removal

To overhaul the carburetor, you have to remove it from the engine. This is a straightforward process that starts by taking off the air cleaner; disconnecting the gas line, accelerator linkage, and choke cable; and then unbolting the carburetor from the intake manifold.

To remove the gas line, you will need two open-end wrenches. The wrench sizes are usually 1/2in and 9/16in. One wrench will be used to hold the carburetor inlet fitting while you loosen the gas line fitting with the other. Adjustable wrenches should not be used because they can easily slip off the soft brass fittings, rounding the edges.

The throttle linkage may be held in place by tiny clips. If so, be careful not to break or lose these fasteners. The choke cable usually pulls free after you loosen a clamp and setscrew.

When removing the carburetor, be very careful not to drop the carburetor nuts or lock washers down the hole in the intake manifold. As soon as the carburetor is off the engine, plug the inlet hole in the manifold with a clean rag to prevent dust or stray parts from entering.

If the engine has been operated recently, the float bowl will still contain gas, most of which can be poured into a suitable container by turning the carburetor on its side.

Carburetor Rebuild Kits

Before taking the carburetor apart, you need to locate a new set of gaskets. Carburetor gaskets are included in what is called a carburetor kit. Actually two types of carburetor kits are available. One is intended for a complete rebuild and, in addition to gaskets, will contain a new float, a needle and seat, an accelerator pump, and numerous small parts such as the check ball, springs, clips, and fasteners. The other, called a jiffy kit, will include gaskets plus a few basic tune-up parts. For a carburetor rebuild, you should have the complete kit.

To cover a broad range of model applications, some rebuild kits include extra gaskets and small parts not needed for your repair job. To make sure you use the right parts, you will sometimes need to compare old parts from the carburetor with new parts from the kit. Often the rebuild kit will include an instruction sheet for servicing the carburetor. This sheet is useful but should be considered a supplement to the carburetor overhaul instructions in the service manual.

Just because the carburetor is sitting atop a thirty-year-old engine is no reason to think that the rebuild kit will be hard to find. The first place to look is a local auto parts store. I have found NAPA parts stores to be unbeatable in their ability to supply parts I would have expected to be long out-of-stock. In many cases, you will find that a local NAPA auto parts store has parts for a thirty- or even forty-year-old Ford truck on the shelf. If not, it can often have the parts delivered the next day.

When the auto parts store does not have, and cannot order, a carburetor kit for your truck, the next step is to contact a specialty supplier or a vender that specializes in carburetor repair. Before contacting the supplier, find out the carburetor's manufacturer and model number and write down this data so that you have it available when you communicate with the vender. This information is usually stamped on the metal tag attached to the float bowl and may be embossed on the carburetor casting. If the tag is missing—which probably means it was discarded in an earlier rebuild—and no information can be read from the casting, the make and model will be listed in the service manual for your truck.

Carburetor Disassembly

Taking a carburetor apart is a simple process. Basically the carburetor casting consists of three parts: the air horn assembly at the top; the body, which contains the float bowl; and the flange base, which holds the throttle valve and is sometimes cast integral with the body. Screws hold the air horn and the flange base to the body.

The choke linkage, which runs up the side of the carburetor, will have to be disconnected in order to

The top of the carburetor consists of the air horn assembly, which contains the choke butterfly and shaft. One of the steps in checking the condition of the carburetor is to inspect for play or a bend in the choke shaft. Nathan Brownell

A carburetor is really a simple fuel-metering device and can be rebuilt by the average weekend mechanic using common shop tools. Rebuild kits for vintage Ford carburetors are readily obtainable, often available from a local auto parts store. Nathan Brownell

The base of the carburetor holds the throttle valve or valves (two on a two-barrel or dual-throat carburetor). The throttle valve shaft also should be checked for wear or a possible bend that may cause the valve or valves to bind in a partially or full-open position—a potentially dangerous condition. Nathan Brownell

separate the casting elements. Before undoing the linkage, take a close look at how it functions and fits together. If you have a service manual, when reassembly time comes, you may be able to refer to photos showing the linkage. If you lack a manual or a clear view of the linkage, it is a good idea to take several photos of the carburetor with the linkage intact, or draw a schematic of how the pieces fit together, for future reference.

While the carburetor is still in one piece, check the choke and throttle shafts for wear, which is felt as slop or play where the shaft passes through the casting. The choke is the large valve at the top of the air horn. The throttle is located in the flange base casting. On some very early carburetors, the choke and throttle shafts are made of brass. More commonly, though, the shafts are made of soft steel. Both the choke and throttle shafts turn in holes drilled through the carburetor casting, another soft metal. In normal operation, wear occurs on both the shaft and the casting. The shaft wears undersize where it passes through the carburetor casting, and the holes in the casting wear in an oval shape.

Because the throttle is in relatively constant movement while the engine is running, wear is far more likely to be found on the throttle shaft than on the choke shaft, which moves mainly when the engine is being started. Signs of a worn throttle shaft are rough idling and a whistling sound sometimes heard on acceleration. If the throttle shaft is worn enough to slop around in the casting, a less worn carburetor should be used.

If the throttle shaft is loose and a carburetor body with less wear cannot be located, you'll need to consider resizing the shaft. The simplest way to do this is to locate a slightly oversize shaft and rebore the holes in the carburetor base to fit. Although this repair is relatively simple, it is usually impractical because finding a new shaft for most older carburetors is nearly impossible. This means that if the throttle shaft and casting holes are worn enough to need fixing, you'll probably have to send the carburetor away to a repair service.

To take the carburetor apart, simply remove the screws holding the air horn and flange base to the body. As these parts are separated, you'll also have to unhook portions of the choke linkage—a process that can be something like working a Chinese puzzle. When the air horn is removed, you will see the float and nearby, in what appears to be a narrow well, another mechanism that looks like a plunger. This is the accelerator pump that provides the engine's fuel needs when you stomp down on the throttle. At the bottom of the float bowl, you'll also see what appears to be a copper screw. This is a jet that passes a metered amount of gasoline into the air stream drawn down through the carburetor and into the engine.

Carburetor Cleaning

Before cleaning the carburetor of the grease and grime it has accumulated from its years of service, first remove the float, accelerator pump, and other internal parts. The float is removed by slipping out the thin copper pin that acts as a hinge. Always check the float for leaks by shaking it beside your ear. If a slight sloshing sound is heard, the float has a pinhole leak in a seam or other spot, that is letting gasoline seep into the hollow interior. A leaking float will raise the gasoline level in the float bowl, resulting in a too-rich fuel mixture and possibly flooding the engine. If the leak allows the float to become filled with gasoline, the carburetor could overflow, sending a stream of gasoline down onto the intake manifold. This is a dangerous situation.

The accelerator pump is removed by disconnecting the plunger from its linkage.

Other parts—the main jet, check ball, fuel-metering rod, and idle-adjusting screw (located in the flange base)—should also be removed prior to cleaning. Instructions in the carburetor kit should show the locations of these parts; if not, they will be

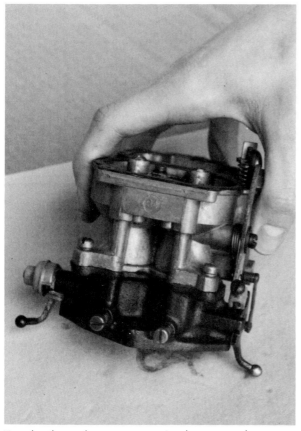

To take the carburetor apart, simply remove the screws holding the air horn and flange base to the body. (In this photo, the air horn has been removed.) As the parts are separated, you will also have to unhook or disconnect the choke linkage. Nathan Brownell

identified in an exploded diagram of the carburetor found in the service manual.

Several methods can be used to clean a dirty carburetor. A recommended noncaustic approach is to use an industrial-strength detergent such as trisodium phosphate, which can be purchased at most discount marts. The detergent is mixed with hot water, and the carburetor housing is placed in the cleaning tank for several hours. The parts are then washed with clear water. Any remaining grime can be scoured off with a vegetable brush or toothbrush.

Another cleaning method is to soak the carburetor castings in mineral spirits to remove grease and grime, then clean the outer and inner surfaces with carburetor cleaner or a strong solvent like lacquer thinner. Both carburetor cleaner and lacquer thinner are highly flammable and should be used with caution and disposed of properly. You will probably purchase the carburetor cleaner in a spray can, since a quantity sufficient to immerse the castings is moderately expensive. If the spray can approach is used, give the castings a good soaking with cleaner, allow the solvent to work a few minutes, then scrub caked-on grime with a parts-cleaning brush. Wear rubber gloves for this operation, since carburetor cleaner is a strong solvent. The housing is rinsed with clean water after each scrubbing. If any gasket material remains on the mating surfaces, these surfaces need to be scraped clean with a single-edged razor blade or Xacto knife.

When the carburetor is thoroughly cleaned outside, spray carburetor cleaner through all holes and openings, then visually inspect these passages to make sure they contain no build-up or residue. Finally blow all passages clean with an aerosol air gun of the type used to clean photographic equipment—available in photo supply stores—or a needle-tip blow gun from an air compressor, to make sure no obstructions or residue remains.

Throttle and Choke Shaft Inspection

Before starting to rebuild the carburetor, check the throttle and choke assemblies for bent shafts. Bent shafts are not all that uncommon on a carburetor that has seen much use and some abuse.

When the air horn is removed, you will see the float, lower left, and another mechanism (in what looks like a narrow well) that resembles a plunger, top center. This is the accelerator pump, which provides the engine's fuel needs when you press down on the accelerator. Nathan Brownell

A bent choke shaft can be caused by a mechanic sticking a screwdriver down the air horn to hold the choke open. A bent throttle shaft is likely to be caused by a severe backfire or someone jamming the gas pedal to the floor, causing the linkage to kink the shaft on that end. Signs of a bent choke or throttle shaft are failure of the butterfly valve to seat properly against the housing or a visible kink in the portion of the shaft that sticks out of the housing.

Often a bent choke or throttle shaft can be straightened without removing the shaft from the carburetor housing. This operation is somewhat delicate and requires care. If the shaft is bent inside the housing, tap against the bend with a light hammer and brass drift until the butterfly valve seats evenly. If the bend is outside the housing, the shaft may be straightened with pliers. Wrap a piece of cloth or paper towel around the end of the shaft to prevent the teeth of the pliers from chewing into the shaft.

If the shaft has to be removed to be straightened, the first step is to turn out the screws holding the butterfly valve to the shaft. Sometimes the ends of these screws have been peened over to prevent them from loosening. If this is the case, remove the peened area with a small file, then turn out the screws.

Before removing the butterfly valve, note that the lip that closes down on the housing is beveled on the bottom side, whereas the lip that closes up on the housing is beveled on the side facing you. To seat properly, the valve will have to be reinstalled with the bevels in this same orientation. So that you do not mistakenly reverse the orientation of the valve—with the result that both bevels would face the wrong way and the valve would not shut off the airflow—mark the top of the valve with a piece of chalk or a marker pen.

Sometimes the valve can be slipped out of the slot in the shaft by hand. If not, it can be pulled loose with a pliers. Often it will have to be rocked back and forth slightly in the shaft to be pulled free.

With the valve removed, the shaft will slide out of the housing. Now you can straighten the bent shaft by rolling it on a flat surface and tapping against the bend with a brass drift.

When the shaft is ready to be reinstalled, coat it with light grease—Vaseline works well—so that it will slip easily through the holes in the carburetor housing. Likewise coat the butterfly valve so that it will slide easily into the slot in the shaft. It may be necessary to reposition the valve in the shaft until it seats evenly on the carburetor housing. Remember to orient the valve with the beveled edges facing the housing. To check for proper fit, close the valve and hold the carburetor housing to the light. If no light can be seen around the edges of the valve, it is seating correctly. Continue to hold the valve closed, and replace the screws that attach the valve to the shaft. Either peen the screw ends or place a drop of Loctite on the threads, to prevent the screws from loosening and dropping into the intake manifold. If you decide to peen the threads, be sure to support the shaft while tapping on the screw ends.

Carburetor Rebuild

For a carburetor to work properly, it is critical that the mating surfaces of the needle and seat be clean and smooth. Because these tight-fitting parts can become worn, the carburetor kit typically contains replacements. The needle and seat control

Before you clean the carburetor, the float, accelerator pump, and other internal parts should be removed. A plastic food container makes a handy storage place for the float, screws, and other small parts.

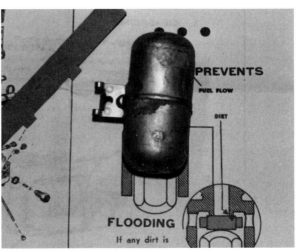

The last step inside the carburetor is to replace and set the float. The float must be set to ride at the correct level. Most carburetor kits contain a paper gauge used to measure the correct float setting. This gauge consists simply of a strip of heavy-stock paper with two tabs on one side.

fuel flow and are located behind the float. The seat threads into the outlet hole on the float bowl and becomes the attaching point for the fuel line. Since the needle—the cone-shaped part—is held in place by the float on many carburetors, you won't slip it into the seat until just before replacing the float.

On carburetors from vehicles that have been parked for a number of years, the accelerator pump will have dried out, causing the engine to sputter and spit when the gas pedal is tromped for fast acceleration. To remedy this, the repair kit also includes a new accelerator pump, which is the plunger that fits into the well beside the float bowl, as well as a replacement for the check valve at the bottom of the well, which is often nothing more than a clip and a tiny steel ball about the size of a BB. The check valve prevents fuel from draining out of the well, which would cause hesitation on acceleration and raise the gasoline level in the float bowl.

When the accelerator pump is replaced, the spring and arm are removed from the old pump and fitted to the new. Then the pump can be slipped back into the well, and the arm fed through the hole that exits at the bottom of the casting. Later this arm will be connected to the throttle linkage.

A carburetor rebuild kit should show the procedure for replacing the check valve and accelerator pump as well as the main jet, metering rod, and other internal parts. If new springs and gaskets are supplied with the kit, substitute these for the originals. Some adjustment may be called for when replacing the metering rod. If so, these instructions should also be contained in the kit's procedure sheet or the service manual.

The last step inside the carburetor is to replace and set the float. This procedure is critical to the carburetor's proper operation and must be done accurately and correctly. Remember that if the needle fits into the seat from the bowl side, it must be inserted before the float is replaced. Wipe the needle with a dry, soft paper towel before slipping it into the seat.

It is essential that the float be installed right-side-up. When you look at the float, no side may appear to be the top, but the orientation of the hinge gives the float a top and a bottom. Normally the loops for the hinge face the bottom of the float bowl. If in doubt, the kit instruction sheet or parts diagram from the service manual should show the proper float orientation. Now the float can be slipped into the float bowl and held in place by inserting the pin through the hinge.

The float must be set to ride at the correct level. Most carburetor kits will supply a paper gauge to use in setting the float. If a gauge is not included, measurements for the proper float setting can be found in the specification sheet with the kit or the service manual. Typically the gauge with the carburetor kit will look like a short ruler with two tabs protruding on one side. This type of gauge is placed across the top of the float chamber so that the tabs hang down toward the float. In the raised position, the float should just touch the tabs.

The float's purpose is to control the fuel level in the float bowl. As the fuel is drawn into the carburetor's air stream, the float drops. When the float drops low enough to release the needle, more gasoline flows into the bowl. This raises the float and cuts off the fuel supply. The float is adjusted to the correct height by bending the arm or tab on the hinge slightly until the top of the float reaches the required height. This is a simple procedure, but one that demands accuracy.

Before reassembling the carburetor, make sure the float drops easily and does not bind on either the hinge or the needle. To check the float action, tip the air horn or bowl assembly—whichever the float is attached to—back and forth in your hand, watching the movement of the float. Old gaskets, found in carburetor kits that have been on a supplier's shelf for many years, have usually shrunk from their

The float's function is to control the fuel level in the float bowl. Before reassembling the carburetor, make sure the float drops easily and does not bind on its hinge. Nathan Brownell

correct size. Assembling the carburetor with these undersize gaskets can also cause the float to bind. To make the gasket pliable so that it can be stretched to its original shape, soak it in water for an hour or so before installing.

In addition to making sure the gasket between the body and the air horn fits the outline of the float bowl and does not interfere with the operation of the float, also check the gaskets between the body and flange base to see that all holes align properly.

Warning: Never discard any parts until you are finished with the rebuild and are sure the carburetor operates properly. In the event that the flange base gaskets supplied with the kit are for a different application, for example, you can—if necessary—reuse the old gaskets.

Now you will insert and tighten the screws holding the air horn and flange base—or base and spacer—to the body. With the carburetor reassembled, you're faced with putting together that Chinese puzzle: the linkage. Chances are, you'll need to refer to the photos or diagrams you made before disassembly, or to an illustration in the service manual, as a guide in putting the linkage together.

If you are not going to replace the rebuilt carburetor on the engine right away, seal it in a plastic bag to keep out dust and moisture. Nathan Brownell

Carburetor Inspection and Reinstallation

With the carburetor reassembled and ready to mount back on the engine, make one more inspection to be sure the float moves easily, the linkage works correctly and engages the accelerator pump when the throttle is opened, and the throttle and choke shafts turn smoothly without binding. To check movement of the float, listen for the clicking sound of the float releasing and closing the needle as you rock the carburetor back and forth in your hand.

If everything appears to work correctly, the carburetor is ready to be reinstalled on the engine. To do this, fit the base gasket over the studs on the intake manifold, slide the carburetor flange down on the studs, and replace the lock washers and nuts. Next you can reconnect the throttle linkage and choke cable—assuming the carburetor has a manual choke. When connecting the throttle linkage, make sure the throttle closes all the way with the accelerator pedal released. If necessary, readjust the linkage where it connects to the throttle shaft.

Carburetor Adjustment

Since the idle screw was removed for cleaning, it will have to be adjusted. To set the idle screw so that the engine will run, turn it between one half and one and one half turns off the seat. If the throttle screw has not been tampered with in the rebuild, it can be left where it is and the engine should run. If the throttle screw has been removed and replaced, turn it in enough to keep the engine running. Final adjustments can be made under the hood. As you fine-tune the carburetor, you're trying to get the highest vacuum or revolutions per minute (rpm) at idle. To do this, you will adjust the throttle and idle settings in tandem, often turning in the throttle screw slightly, then backing off the idle, and so forth, until the best setting is reached.

Air Cleaner Servicing

The last step in overhauling the carburetor is servicing the air cleaner. Old-style oil bath air cleaners should be washed out and refilled with clean oil; the owners manual for your truck may say

---

How to Check for Carburetor Air Leaks

The sign of an air leak when the vehicle is running is a rough or fast idle. To check for air leaks at the flange base gasket, around the throttle shaft, or at the intake manifold gaskets, spray carburetor cleaner around these areas with the engine running. If the engine speed picks up, you have an air leak. If the air leak is around the manifold or carburetor base, the gaskets in these areas should be replaced. Most carburetors that have seen a lot of use leak some air around the throttle shaft.

to do this each time the engine is serviced. Dry wire mesh air cleaners should also be washed with solvent and blown dry with an air hose. A light coating of oil sprayed into the mesh increases this type of filter's effectiveness. Modern paper filters are simply replaced.

A carburetor overhaul can make a dramatic difference in an engine's responsiveness, economy, and overall performance. The process is simple enough for any old-truck owner to perform successfully, and applies with little increase in complexity to two- and four-barrel carburetors as well.

Rebuilding the Generator and Starter

Although they're actually part of the electrical system, the generator and starter attach to the engine and are included in this engine overhaul chapter, since rebuilding these two components is more mechanical than electrical in nature.

Although one creates electrical power and the other uses electrical power, the generator and starter are mechanically similar devices and share many of the same parts. Given the recommended periodic maintenance, these electrical accessories were built to last the life of the truck. But because previous owners probably overlooked applying a few drops of oil to the armature bushings every six months or so—as called for in the owners manual—and neglected to replace the brushes at 50,000-mile intervals, or thereabouts, it's likely that the generator and starter on your truck are candidates for cleaning, inspection, and overhaul. If your truck is undergoing restoration, these components, like everything else on the vehicle, will be completely disassembled and rebuilt.

Signs of Generator Problems

Even though generators and starters are mechanically similar, their functional differences require that they be diagnosed and overhauled separately.

Generator problems are of three types: no charge, too much charge, or not enough charge. Usually if the generator is not putting out current, the problem is with the generator itself. If it's putting out too much or too little charge, the problem is likely to be with the voltage regulator or the wiring.

Assuming the truck runs and the wiring to the ammeter is good, you'll know the generator isn't

Even though one creates electrical power and the other uses electrical power, the generator and starter are mechanically similar devices that share many of the same parts. A starter is shown here.

working if the ammeter's needle doesn't move into the "+" zone—the charging zone. When this occurs, first check that the generator belt is tight and is turning the generator pulley. If the belt is OK and the ammeter works—shows discharge when the lights are turned on—a few simple tests will determine if the charging problem lies with the generator.

First make sure that all electrical connections are in good condition—it's important that the connections be clean and tight—and that you have a good battery in the system. On early trucks using a cutout-type regulator, charging problems can occur in either the generator or the cutout. To isolate the cutout, put a jumper wire between the output lead of the generator and the battery terminal on the cutout. With the engine running at fast idle, a generator that is functioning normally will show a charge rate on the ammeter. Be sure to disconnect the jumper wire when you shut off the engine, to prevent the battery from discharging through the generator.

On the later systems using a conventional voltage regulator, problems can also occur in either the generator or the regulator. To isolate the regulator, disconnect the field wire at the F-terminal of the regulator and reconnect it to the A-terminal, or armature terminal, on the regulator. Now run the engine at fast idle, and a good generator will produce current flow, which should be indicated by the ammeter. If the ammeter shows that the generator is charging, the problem is usually at the regulator. If the generator does not charge, the problem is most likely in the generator, but the regulator could still have problems.

Run the engine for only a short period of time when doing this test, because removing the regulator causes the generator to give full electrical output and it will quickly overheat.

Generator problems are usually traceable to worn or sticking brushes, a dirty or worn commutator (the smaller-diameter copper section against which the brushes ride), worn bearings, or an internal short. An overhaul covers all of these problems.

Generator Removal

To overhaul the generator, you have to remove it from the engine. Thankfully the engines on older vehicles aren't covered with a tangle of hoses. Before removing the generator, be sure to disconnect the ground cable to the battery. Now remove the wires that connect to the armature, field coil, and ground lugs. Unless the wiring will be replaced, mark each wire with its attaching point on the generator as it is removed.

Now loosen the bracket bolt that holds the generator pulley tight against the fan belt. This allows the generator to be pushed against the engine and the belt to be slipped loose of the generator pulley.

The generator is supported by a mounting bracket that bolts to the engine block. This bracket is attached to the underside of the generator. To gain access to this bracket, it is best to swing the generator as far away from the engine as it will go. This should allow you to reach the bolt that holds the generator to its mounting bracket. With this bolt removed, the generator can be lifted out of the engine compartment.

Generator Disassembly

To open up the generator, simply remove the cover band, if present, and the bolts that hold the end frames to the housing. The armature, with the front end frame and pulley attached, can now be pulled out of the housing. If you are rebuilding the generator—as opposed just to inspecting and cleaning it—this is a good time to pull the brushes out of their holders and disconnect their wire leads from the brush end frame.

Normally this is as far as the generator needs to be disassembled. Removing the field coils and pole shoes (the large internal parts that line the inside of the housing) is difficult because the screws that hold these parts in place are often hard to loosen and usually require a special tool. These parts may also be damaged by removing them. Unless they are defective, the field coils and pole shoes are better left in place. This is also true for the generator pulley. Unless the front bearing is defective, the pulley and end frame need not be removed from the armature. A worn bearing should have made itself known by a squealing or chattering sound when you tested the generator.

If the earlier tests showed the generator to be defective, the field coils need to be tested for a

To open up either the starter or the generator, simply remove the cover band, if one is present, and the bolts that hold the end frames to the housing. The armature, left, can now be pulled out of the housing, right.

possible short. First, though, you will want to clean the generator and its parts.

Generator Cleaning

Usually the generator is covered with caked-on dirt and grease. This buildup should be scraped off external surfaces with a putty knife before they are washed with parts cleaner. The housing, end frames, armature, cover band, and generator pulley can now be washed in solvent. Since the field coils have not been removed, you should not immerse the housing in the solvent. Doing so might damage the insulating wrapping on the coils.

You can use a parts cleaner brush to remove grease that may have collected inside the housing. Likewise you will use the parts cleaner brush to get at the grease that has accumulated on the back side of the pulley and front end frame. Gunk degreaser is effective for cleaning areas like this. This product, which is available in the automotive department of discount marts as well as in automotive parts stores, comes in an aerosol can and is sprayed onto the grease-coated surface, allowed to work for a few minutes, then simply washed off with water from a garden hose. Stubborn grease coatings may require a couple of applications.

If the exterior of the housing is rusted, it can be sanded or cleaned using an acid jell like naval jelly or Jenolite.

Generator Internal Testing

If the generator wouldn't put out current when you ran the earlier diagnostic test, you will need to run another set of tests now that the unit is disassembled, to locate the problem. These tests will check for open or grounded circuits in the field coils, armature, and brushes. If the generator passes all these tests, new brushes and a cleaned commutator should restore it to working order.

To do the internal tests, you will need a volt-ohmmeter (VOM). This electrical test device can be purchased at electronics stores like Radio Shack for under $25. If you have not used this test device before, read the instructions to become familiar with how the meter works.

Field Coil Testing

Two tests are performed on the field coils. The first checks for an open circuit. With the VOM indicator on the ohmmeter setting, place one lead from the meter on the field coil terminal (the insulated bolt that sticks out of the generator frame) and the other lead on the field coil that attaches to the brushes. If the needle on the meter does not move, the field coils have an open circuit—electricity is not passing through. An open circuit requires that one or both of the field coils be replaced. Separating the field coils electrically and checking each should pinpoint the source of the problem.

The second test checks for a ground between the field coils and the housing. For this test, place one lead from the VOM on the field coil terminal and the other on the generator housing. Make sure the field coil terminal that attaches to the insulated brush holder is disconnected and does not touch the housing. If the needle moves, the field coils are shorted to the housing. To verify that the problem is

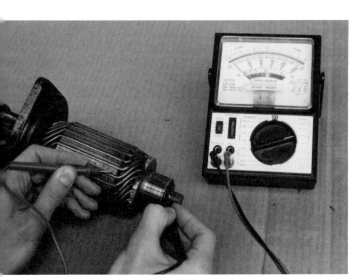

To check for a ground inside the armature, place one test lead on the armature core and the other on the commutator (the smaller-diameter section on which the brushes ride). If the needle in the VOM moves, a short exists in the armature and this part will have to be replaced. Nathan Brownell

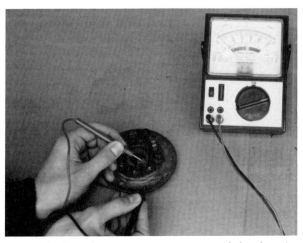

As you look at the starter or generator end that has the brush holder brackets, you will notice that one set of brackets is insulated from the end frame. To check for a short in this insulation, place one lead from the VOM on the brush holder, and the other on the end frame. If the needle moves, the insulation is allowing current to pass and the end frame will need to be replaced. Nathan Brownell

not at the field terminal, disconnect the field terminal from the housing and try the test again. If the needle does not move this time, replace the insulating washers on the field terminal, and this should correct the problem.

Armature Testing

With the armature, again two tests are involved. The first checks for a short in the armature terminal. This is done by placing one lead from the VOM on the armature terminal and the other on the generator housing. If the needle moves, a short exists and the insulating washers on the terminal should be replaced.

The second test checks for a ground inside the armature. This is done by placing one test lead on the armature core or shaft and the other on the commutator. If the needle moves, a short exists in the armature and this part will have to be replaced.

If the commutator is dirty (doesn't show a bright copper finish), it should be cleaned. This is done by spraying on carburetor cleaner and wiping with a paper towel. Several applications of cleaner may be necessary to restore the copper sheen. If the commutator has a worn or rough feel, or if one or more segments appear to be burned, it will need to be turned and the mica (the separators between the copper segments) undercut. This can be done at a generator or starter repair shop. A more serious problem is solder melting out of the winding-to-commutator connections owing to overheating sometime in the generator's past. If this has happened, you will also see solder deposits on the inside of the housing near the brush holders. If overheating has caused some of the armature windings to separate from the commutator, the armature should be replaced.

Brush Holder Testing

As you look at the end of the generator that has brush holder brackets, you will notice that one set of brackets is insulated from the end frame. To check for a short in this insulation, place one lead from the VOM on the brush holder and the other on the end frame. If the needle moves, a short exists in the insulation.

Generator Painting

Painting is best done while the generator is still apart. Be sure to mask the nameplate and terminals. For a quality finish, prime and paint using a professional-style spray gun.

Generator Reassembly

As you reassemble the generator, also inspect and replace the bearing and bushing, if necessary. If the bearing on the end frame with the pulley is worn, you should have heard a howl, whine, or chatter during the tests when the generator was still hooked up to the engine. Discoloration due to heat or a rough feel when the bearing is turned are other signs that this bearing should be replaced. To replace the bearing, you will have to remove the drive pulley. This will require a puller.

The bushing in the end frame that has the brush holder is likely to be worn, particularly if previous owners forgot to squirt a few drops of oil onto it from time to time. You can check for bushing wear by inserting the armature shaft into the bushing and measuring the clearance. If more than a 0.004in gap exists, the bushing should be replaced.

Bushings and bearings are available from vintage Ford parts suppliers like Dennis Carpenter Ford Reproductions and can often be purchased at the local NAPA parts store. When looking for bushings or bearings at a parts store, you will need to have the width as well as the inside and outside diameters from the old parts. These measurements should be taken with a micrometer. If you lack this tool for making precision measurements, just take the old bearing or bushing to the parts store and the counter clerk should be able to take the measurements.

When these trucks were in regular service, rather than replace the bushing, repair shops just installed a new end frame. You might ask the parts store clerk if a new end frame for your truck's generator is in stock. Probably the clerk will laugh, but it just might be, and if so, you will be saved the job of replacing the bushing.

Plan to buy new brushes when picking up the bearing, bushing, or end frame.

The easiest way to remove the old bushing is to turn it out with a tap. First, though, you will need to find a washer that will fit inside the bushing. The washer's purpose is to support the tap. Just drop the washer into the bushing hole, and then select a tap large enough to thread into the bushing. Insert this

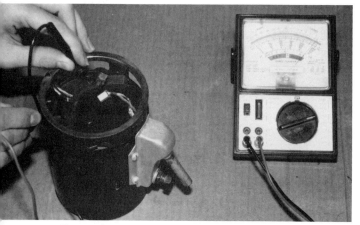

To test for an open circuit in the starter field coils, place one lead from the VOM on each field coil lead (these are the wires that attach to the brush holder when the starter is assembled). The meter should show current flow. Nathan Brownell

tap into the bushing and turn it until it seats against the washer at the bottom of the bushing hole. Continue to turn the tap, and the bushing will climb up out of the hole. The new bushing should be installed with a driver. Lacking this, you can use a socket.

The next step is to reconnect the leads from the field coils and insert the brushes into their holders on the end frame. The clips that hold the brushes against the armature when the generator is operating will have to be pulled back in order to insert the brushes into the holders. You can pull the clips back with needle-nose pliers, but a better tool can be made by cutting about a 4in piece of coat hanger wire and bending one end into a hook. Make a 90-degree bend about 1in from the other end to form a little handle to use in pulling against the spring's tension.

Instead of letting the clips snap back against the brushes, carefully rest each clip against the side of the brush that is just started into the holder. With the clips not pressing on the brushes, you can slide the armature through the housing and into the brush end frame. Before inserting the armature, smear a dab of high-temperature grease onto the end cap bushing. This will ensure that this bushing is well lubricated for many, many miles. With the end plates tight against the housing, you can replace the two bolts to hold the parts together. Now push the brushes down against the commutator and move the clips onto the brushes, then replace the cover band if the generator has one, and the reassembly is complete.

Generator Polarizing

When a generator has been off the vehicle for testing or repair, it may lose its residual magnetism. Its magnetism will need to be restored to the correct polarity before the generator will charge. This must be done before the engine is operated. To polarize the generator, disconnect the wire to the field wire at the regulator terminal and momentarily connect it to the battery terminal.

Generator Bench Testing

Before reinstalling the generator on the truck, it's a good idea to test it for current draw. In this test, you will be operating the generator as a motor, and so it will be drawing, not producing, current.

To motorize the generator, connect the battery and an ammeter in series between the generator armature terminal and the proper battery post—the positive post on negative ground systems, the negative post on positive ground systems. Ground the field terminal to the generator housing with a jumper wire. Attach a ground cable to the generator housing and the other battery post, and the generator should operate as a motor—it should spin.

On most generators, the ammeter should register between 4 amperes (amps) and 6amps. If the current draw is not within this range, check for poor connections. If the connections are good, the problem could be a faulty armature or field coils, or possibly a poor brush fit on the armature. If you suspect that the brushes have not worn to the curvature of the armature, you may want to run the generator as a motor for a short period to let the brushes "wear in." Then give the generator time to cool to room temperature, and repeat the test again to check the current draw. If it is within the acceptable range, the generator is ready to be installed on the truck.

This motorizing test also polarizes the generator.

Generator Reinstallation

After replacing the mounting bolts at the bottom of the generator, slip the belt over the generator pulley and pull the generator tight on the belt. Now screw in the bolt on the generator bracket and check the belt for correct tightness—7/16in to 1/2in deflection of the belt at a point midway between the fan and generator pulleys. Retighten or loosen the generator as necessary to achieve this tightness. Finally reconnect the wiring leads to the generator, and this accessory should be in sound operating condition.

Signs of Starter Problems

Starter problems are noticed when the starter turns slowly, if at all, on a well-charged battery. When poor starter action occurs, you will want to determine whether the cause of the problem lies in the wiring or the starter. A simple way to do this is to turn on the headlights and crank the starter.

If the lights go out, the battery voltage is too weak to operate the lights and starter motor. This can be caused by a run-down battery or bad electrical connections or cables. A poor connection almost always can be traced to corrosion on the battery terminals or a loose cable clamp. Either condition is corrected by removing the battery cables; cleaning the posts and cable clamps with a solution of baking soda and water—be very careful that none of this solution runs into the battery cells; and then scrubbing the posts and cable clamp holes with a battery post brush. To prevent the terminals from corroding again, replace and tighten the cable clamps, then spray the clamps and terminals with Plastidip, a spray-on plastic coating available in auto supply stores and discount marts. After making sure the battery has a good charge, try this test again. In most cases, the starter will now spin normally.

If the lights dim so that they can barely be seen and the starter turns very slowly or not at all, either the battery is in a run-down condition or some other mechanical problem is throwing a very heavy load on the starter. The battery's condition can be checked easily with a hydrometer. Lacking this inexpensive testing device, you can put a charger on the battery and give it a charge. If it takes the charge, bring the battery to a fully charged state and try this test again. If the problem was a run-down battery, the starter should spin normally.

Mechanical problems that could cause the starter to turn sluggishly or not at all are water in the cylinders, a bent rod, seized piston rings, and seized bearings—to name the most common. If the starter itself is at fault, possible problems include a bent armature, loose starter bolts, or worn bearings. Of these, on an older truck, worn starter bearings is the most common.

If the lights stay bright and the starter won't crank, an open circuit exists—in the starter, the starter switch, or the control circuit. If the starter is equipped with a solenoid, this control device can be bypassed to remove it from the circuit. The entire starter control circuit can be bypassed by connecting a heavy jumper—a battery jumper cable works well—directly from the hot battery post to the starter terminal. If the starter turns over, the problem is in the control circuit, which will have to be checked out component by component. If the starter does not operate by the jumper, either it does not have a good ground or it has an internal mechanical problem. To check for a bad ground, first make sure the starter is bolted tight against the bell housing. When the bolts are tight, clamp one end of a jumper cable to the starter housing and the other end to the ground terminal on the battery. If the starter still won't respond, it definitely has a mechanical problem and will need to be removed from the truck for repair.

Starter Removal

Always disconnect the ground cable from the battery before working on any electrical component. To remove the starter, loosen and remove the large brass nut that attaches the heavy cable to the starter. If the solenoid is mounted to the starter housing, remove the small wires to the solenoid and label them. Now you can loosen the bolts that hold the starter to the bell housing. Some starter bolts are accessed best from underneath, and the starter can generally be removed easiest from this direction. If you find any shims between the starter and bell housing, note their position and save them. They will be needed when you replace the starter.

Starter Disassembly and Cleaning

A starter comes apart in basically the same way as a generator, and the steps for cleaning a starter are also the same as those for cleaning a generator.

Starter Testing

Inspect and test the armature following the steps described under the "Armature Testing" subheading in the "Rebuilding the Generator and Starter" section earlier in this chapter. If the armature is bent or shorted, it should be replaced. If the armature tests OK, clean the commutator. If the commutator is rough or out-of-round, have it cut smooth at a machine shop.

Also inspect the starter terminal (this is the copper bolt to which the starter cable attaches) for stripped threads. If the threads are damaged, this bolt should be replaced. If the bolt is replaced, it will be necessary to resolder the connections to the field coils.

Next test the field coils, following a procedure similar to that used to test the generator field coils. To test for an open circuit, place one lead from the VOM meter or a test light on each field coil lead (the field coil leads are the wires that attach to the brush holder when the starter is assembled). If the meter does not show current flow or if the test light does not light, the field coils are not making a complete circuit and should be replaced.

To test the field coils for a ground, first make sure the field coil terminals inside the housing are disconnected. Then place one lead from the VOM or a test light on the starter terminal and the other on a clean, unpainted surface of the starter housing. If the meter shows current flow or if the test light lights, a ground exists in the field coil circuit and the coil assembly must be replaced. Be aware that false ground readings may be obtained if the brushes are not all the way out of the brush holder or if the field coils or terminals are touching the housing.

The starter brush holder is tested for ground following the same method described for the generator brush holder.

Starter Painting

The starter is painted using the same procedure as for the generator.

Starter Reassembly

Before reassembling the starter, check the bushing and bearing, and replace them if needed. The procedure is the same as the one described in the "Generator Reassembly" section. Also install new brushes in the brush holder.

If a solenoid mounts to the starter, bolt this component in place and connect the solenoid cable to the lug on the starter frame.

Starter Reinstallation

The starter can now be replaced on the truck in the reverse order of that in which it was taken off. Be sure to match the wires to the correct terminals on the solenoid and tighten the nut holding the starter cable to the terminal post. Assuming that all wiring connections have been made correctly, all the wiring is sound, and the battery is charged, the starter should crank the engine in a lively fashion.

If the generator and starter have been restored for later use, wrap them (a plastic or paper grocery bag works well), label them, and put them aside for later installation.

Engine Rebuild

Now that you have gained some confidence in your mechanical abilities, you may be up to the challenge of tackling the engine. But just because the engine's in a thirty- or forty-year-old truck doesn't necessarily mean it needs an overhaul. It may be a recently rebuilt or low-mileage unit, or just well-cared-for and healthy. So before approaching the engine, wrenches in hand, it's a good idea to run a few tests. From these, you will know whether or not an overhaul is in order, and then you can make a decision about hiring out the rebuild or doing the work yourself.

Engine Assessment

Checking engine compression will give you a fairly accurate reading of the engine's health. The amount of pressure created in the cylinder as the piston comes to the top of its compression stroke tells the condition of the valves (whether or not they are seating well or burned) and piston rings (whether or not the cylinders are worn or scored and if the rings are seating snugly against the cylinders).

To check the compression, you will need a

With the engine removed, your next job will be to clean off years of accumulated grime and grease. The easiest way to do this is with a steam cleaner.

Engine removal is easiest when the front clip (hood, front fenders, grille, and radiator) has been removed. The engine can be lifted with an overhead hoist or an engine hoist, often called a cherry picker.

compression tester—available from auto parts supply stores. The engine section of the shop manual for your year truck should indicate "in-spec" compression readings. Commonly these readings should be between 120psi and 130psi at sea level. If the compression readings are substantially lower or vary more than 5psi to 10psi from cylinder to cylinder, chances are you have already observed the engine burning oil and idling roughly, and perhaps have even heard a sucking sound at the tailpipe—an indication of a burned valve. Write down the compression readings as you take them, because they will help you describe the engine's condition to the rebuilder, if you decide against rebuilding the engine yourself.

Other signs that an engine is in poor health are knocks (indicating worn bearings), low oil pressure (below 10psi once the engine has been warmed up), and oil blow-by (seen as a heavy oil film around the crankcase breather cap). Blow-by is a sign of poorly seating rings or tapered cylinders. If you recognize any of these trouble signs with your engine, a rebuild is in order.

For trucks produced since the late sixties, rebuilt engines are available as either a short-block (the block assembly and internals only, without the oil pan or heads) or a long-block (a complete engine). Often these "factory" rebuilt engines can be purchased at a local auto parts store. You will bring in your old engine for a "core credit." Buying a short-block is typically less expensive than paying for machining work and purchasing new internal parts separately. The advantage of a long-block unit is that it costs less than hiring someone else to rebuild the engine. The disadvantage of replacing the engine in your truck with a factory rebuilt is that the truck will no longer have the original engine. If, however, your main concern is getting your truck on the road in top running condition, these engines are guaranteed to be in-spec and usually the least expensive means to a like-new engine.

Engine Removal

On later Ford V-8 and six-cylinder engines, the head or heads can be removed to overhaul the valves, and crankshaft and rod bearing work can be done with the engine in the truck. In most situations, however, and certainly for restoration, the engine will be removed from the truck. On older vehicles, this is a pleasantly straightforward job that begins by removing the hood and radiator. If the truck is coming apart for restoration, it also helps to remove the grille assembly. Next disconnect all the wiring leads to the engine's electrical components. If new wiring will be installed, it isn't necessary to label the leads, but if the existing harness will be reconnected, you should tag all the leads so that you can later put them back correctly. You will also disconnect the fuel line, carburetor linkage, and choke cable. The exhaust pipe needs to be detached at the manifold; often the bolts here are rust frozen and have to be heated with a torch to prevent breaking.

Now you can loosen the motor mount bolts and remove the bolts that attach the transmission to the engine bell housing. This done, you are ready to pull the engine. The best way to lift the engine is by a sling that attaches to the head bolts. Rigging a sling from a length of heavy chain is not recommended because such a device would not ensure stable control of the engine during lifting. Either an overhead chain fall or a cherry picker engine hoist, available at most rent-all stores, can be used to lift the engine out of the truck.

Engine Cleaning

With the engine removed, the next step is a thorough cleaning. The easiest way to clean off years of accumulated grease and grime is with a steam jenny—a tool that can be rented from most rent-all stores.

The other option is to clean the engine using Gunk (a degreasing cleanser), hot water, a putty knife, and a scrub brush. This approach is time-consuming and tedious.

Engine Disassembly

Once the engine is cleaned, it is ready to be disassembled or hauled to the rebuilder. OK, now you've reached decision time. Are you going to rebuild your truck's engine, or hire the work done by someone else? If someone else is going to do the job,

A handy organizer for engine parts is this plastic tray from The Eastwood Company. It holds all the pistons and connecting rods, in sequence, plus valves and pushrods (on an overhead valve engine). It stores parts from a four-, six-, or eight-cylinder engine.

the next question is, who? Rebuilders for forty-year-old engines don't list themselves in local phone directories. The best advice here is to have your truck's engine overhauled by a vintage Ford engine specialist—not by someone's cousin Bob who's overhauled a bunch of Ford 351 V-8s. In rebuilding an engine, lots can go wrong. If you are a reasonably good mechanic, you have as good a chance of doing the job right yourself as does someone who has worked on some later-model engine and will be learning—maybe not too conscientiously—on your older Ford. If you have enjoyed the mechanical work you've done so far, and are very particular and careful, you have no reason to fear the engine challenge.

Resources

The shop manual for your truck will show the sequence to follow in disassembling and rebuilding the engine, but as mentioned before, it's also helpful to have a *Motor* or *Chilton's* manual on hand, too. If you are tackling an engine rebuild for the first time, you should have someone with experience who you can call on for help or advice. But what's most important is that you pay attention to detail. One thing is sure: If you rebuild your truck's engine yourself, you will experience a surge of pride every time that engine spins to life.

Preparation

Before starting disassembly, gather a bunch of empty coffee cans or plastic milk cartons with the neck cut off, plus several cardboard boxes, to hold bolts and the engine's internal parts. The cans or milk cartons are used to store the bolts for the oil pan, timing cover, and other assemblies so that when you are ready to put the engine back together, you will be able quickly to find the right bolts for each step. Be sure to label each container with its contents. Also clearly mark the pistons and rods with the number of the cylinder they came out of, and on a V-8, the bank. This can be done with a file, punch, or marker. An easy way to keep from mixing up the valves is to drill a series of holes large enough for the valve stem to fit through in a length of wood and place the valves in the holes in the same order that you took them out of the engine. The Eastwood Company sells an inexpensive parts tray that is ideal

If the cylinder taper in your truck's engine exceeds 0.005in, the cylinders should be rebored. Problems associated with cylinder wall taper include rings failing prematurely and power stroke gases blowing into the crankcase.

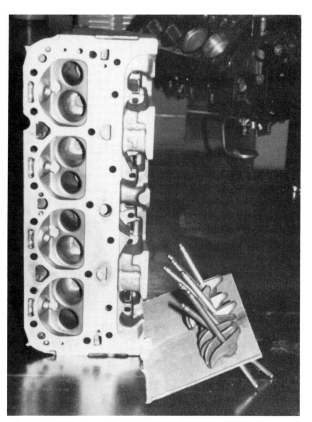

If your truck's engine was built before 1969, ask the machine shop to cut out the valve seats and install hardened seats and guides. The shop will check the valve seat areas for cracks and should be asked to check the head surface that mates with the block, for warpage. If warpage exists, head gasket failure is likely.

A hot rodder's trick for improving engine performance is to polish the combustion chamber and inlet and exhaust ports. This you can do yourself with a die grinder and a cylinder head porting kit. The Eastwood Company

The life of an engine depends on the condition of the crankshaft. The machinist will inspect the crankshaft for out-of-round journals, twists, and cracks.

The crankshaft should also be inspected for wear on the lobes. Some machine shops can regrind camshafts to restore the cam lobe contours.

A crankshaft whose journals have worn below spec can be restored for use, by a machine that builds up the bearing surfaces with welds that are then ground down to specification size.

for organizing all of an engine's internal parts: pistons, rods, lifters, and pushrods.

Tools

To take an engine apart and put it back together, you'll need a few specialty tools. Some of these may be rented or borrowed. For convenience in working on the engine, and to keep from having to wrestle it around the shop floor where it is likely to pick up dust, plan on buying an engine stand. Sturdy yet relatively inexpensive engine stands are available from auto parts stores and discount marts. To take the engine apart, you will need a puller to remove the harmonic balancer and a valve spring compressor to remove and reinstall the valves. To put the engine back together, you will need a ring compressor, brass drift, and torque wrench. This tool list assumes that you will have a machine shop fit the wrist pins in the pistons and perform other specialized procedures.

Before starting the engine, you will want to buy or make an oil pump primer. This tool is used to spin the oil pump to fill the empty oil lines and galleys. You can make an oil pump primer by welding an extra length of metal onto a spare distributor shaft.

Engine Machining

Regardless of who rebuilds your truck's engine, machining work will need to be done. If you're doing the rebuild, part of the job will be finding a reputable machine shop. If friends have had engine machine work done at rebuilding shops in your area, ask them what they thought of the work. Most machine shops are small, and satisfied customers are their best advertisement.

Typically the machinist begins by boiling out the block. This is done to remove sediment that may be clogging the oil passages, and rust scale and other debris that may have settled in the water passages.

After the block is cleaned inside and out, the cylinders are checked for taper (a larger diameter at the top than the bottom) or elongation (an egg shape). Cylinder taper develops as the piston rings rub up and down the cylinder walls. If the taper measures in excess of 0.005in, you should have the cylinders rebored. Since piston sizes run in increments of 0.015in over standard, the cylinders will be bored to match the closest piston size.

Cylinder taper on an older engine often exceeds 0.005in. If the taper in the cylinders of your truck's engine is in the 0.01in to 0.012in range and your purpose is to overhaul the truck to make it serviceable—with cost being a consideration—you could have the machine shop hone the cylinders and knurl the pistons. This will avoid the cost of having the block bored and buying new pistons; you should replace the rings whether or not you are installing new pistons. This combination of honing and knurling can make the engine serviceable by

Grinding a crankshaft is a precision process performed by a skilled artisan. A machinist cannot "touch up" a poorly performed grinding job. If the work is not done right the first time, the crankshaft will have to be reground to the next undersize specification.

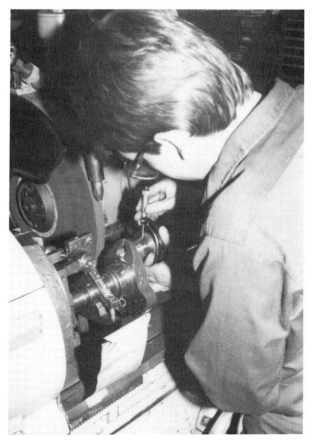

At the completion of each step, the machinist carefully checks the work.

reducing the taper while expanding the pistons.

A good machine shop will tell you whether the taper on your block is severe enough to require that the block be rebored. Problems associated with cylinder wall taper are premature ring failure and power stroke gases blowing into the crankcase.

As part of an engine overhaul, the bores on the main caps are also drilled to the proper size.

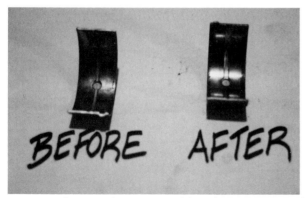

A proper bearing fit is ensured by placing a strip of Plastigage on the bearing insert, then tightening the bearing cap against the crankshaft. This flattens and spreads the Plastigage, which then can be measured for bearing fit.

Severe cylinder taper allows combustion gases to blow past the rings, creating what's called blow-by. You can recognize this blow-by condition in badly worn engines by the smoke that pours out of the crankcase breather cap.

Besides reboring the cylinders, the machine shop will check the mating surfaces on the block and head or heads (a V-8 has two heads) to make sure these surfaces are perfectly flat. If it finds any distortion or warpage, it will mill the block and head surfaces.

Ask the machine shop to magnaflux the head or heads for cracks. Cracks that develop across valve seats are sometimes visible, sometimes not. It is very important that all cracks, visible and invisible, be detected and repaired. Otherwise the engine may leak coolant into the cylinders, resulting in a ruined engine.

In addition, the machine shop can press out the old valve guides and install new guides and seat the valves. If your truck's engine was built before 1969, ask the machine shop to cut out the valve seats and install hardened seats and guides. The additive tetraethyl lead, which was used to make leaded gasoline, provided a lubricating film that kept the soft valve seats found in older, typically pre-1969 engines from wearing. The absence of "lead" in any significant quantities in today's gasoline can lead to premature valve seat wear if hardened valve seats are not installed. This job can only be done with the engine disassembled and out of the truck, so now is the time to do it. The expense is well worth the freedom from concern about future valve seat wear.

The camshaft should also be inspected for wear on the lobes (the triangular extensions off the center of the shaft), which can cause the valves to come only partially open. Some machine shops can regrind the camshaft to restore the cam lobe contours. If this service is not available and the cam is severely worn, the camshaft should be replaced.

If the original rods are to be reused, they should also be checked at the machine shop, for trueness. A bent rod will bind on the crank journal as it revolves, making the engine difficult to turn over initially, and if the engine does start, the bearing in that rod will be short-lived.

The life of an engine and its bearings depends on the condition of the crankshaft. Have the machine shop check the crankshaft for out-of-round journals, cracks, and twists. If any of the journals are found to be out-of-spec, the crank should be reground to the nearest undersize specifications. Also have the crankshaft magnafluxed. If a crack is found, the crankshaft should be replaced.

After grinding, the crankshaft journals are polished and oiled. As a last step, the oil holes are cleaned to remove any debris picked up during the grinding process. From the machine shop's point of view, the crankshaft is now reconditioned and ready to go back into your truck's engine. Before this happens, however, you should take the precaution of measuring each journal with a micrometer. What you're looking for is to see if the journals are ground and polished to exactly the same measurement for the entire length of the journal. If the machinist used a dull grinding stone, it's possible for the ends of the journal to have a slight curve. Should this be the case, you have a real problem. If you assemble the engine with the journals in this condition, the bearings will bind. Unfortunately the machinist can not "touch up" a poorly done grinding job. If the work is not performed right the first time, the crank will have to be reground to the next undersize specifications. This is not at all desirable, and the problems associated with substandard work of this sort are all the more reason to pick a machine shop that has a reputation for doing quality work.

Engine Reassembly

Now you're at the stage where you can make a list of the parts that are needed to complete the rebuild, and set about finding them. Top on your list will be an engine overhaul gasket set. The first place to shop for new gaskets and other needed parts is a local auto parts store. Ford is a popular make and used the same engines for many years in both cars and light trucks, so many mechanical parts are still in production. Items that are not available at an auto parts store can be ordered from a vintage Ford truck parts supplier. You will also want to obtain the correct paint for your truck's engine and decals for the carburetor air cleaner, oil filter, and valve covers (on later overhead valve engines). These supplies will need to be ordered from a vintage Ford parts supplier.

Reassembling the engine is a precision job that requires extreme attention to detail. The steps for this assembly sequence can be found in the Ford shop manual for your vehicle.

Be sure to oil all friction surfaces as you assemble moving parts. This is especially critical for the cylinder walls and engine bearing surfaces. A low-friction oil like STP will give good protection against metal-to-metal contact for the bearings and other moving parts until the oil pump oils the engine prior to start-up.

If you attempt to use an older gasket set (these are sometimes available at bargain prices at swap meets), you may find that the cork gaskets have dried and shrunk. When this is the case, the gaskets should be soaked in water for an hour or so before using to allow them to stretch back to shape. Don't soak paper gaskets.

The best way to ensure a leak-free engine is to lay a thin bead of Permatex Blue gasket sealer on all gasket surfaces—both sides of the gaskets.

When you have finished assembling the block, you can install the oil pan and set this unit aside. Cover the engine block well to prevent dust from settling in the cylinders and other moving parts.

After following the shop manual's instructions for installing the valves, place a fresh head gasket on the block and bolt the head in place. Follow the

A "rodded" Ford flathead engine is a beauty that borders on art. This 59A-B engine is fitted with high-compression heads, dual carburetors, and a "beehive" oil filter as well as internal modifications. Patrick Dykes

Speed equipment for vintage Ford engines can be found quite easily at swap meets. For the flathead-style engines, this consists of high-compression as well as overhead valve heads.

tightening sequence for the head bolts that is shown in the shop manual, and tighten the bolts to the specified torque.

Now the manifolds can be mounted and the engine put back in the truck. This done, the accessories—fuel pump, carburetor, generator, and starter—can be remounted, the fuel lines reconnected, and the wiring harness hooked up to the engine's electrical connections. Use new spark plugs along with new spark plug wires. Be sure to follow an electrical diagram for the engine, so that the wires are inserted in the correct sockets in the distributor cap and connected to the right spark plug.

Engine Testing

Before starting the engine, it's essential to prime the oil pump to lubricate the engine in advance. This is done by either purchasing a primer that is inserted through the distributor hole and spun by a 1/2in electric drill, or making a primer from a scrap distributor. To make a primer, remove the cap, rotor, and any other internals from the scrap distributor, then braze a bolt—with the threads sticking up—to the top of the distributor main shaft. If this distributor shaft has gears that mesh with the camshaft, grind off the teeth using a bench grinder. Now you can insert the shaft into the distributor hole and spin the bolt by hand to check for any binding that would prevent the shaft from turning freely. If you detect no binding, use a 1/2in electric drill to spin the shaft and prime the pump. This will take a few minutes. Air must be expelled from the pump for the lubrication system to become functional. You will feel when the pump begins to work by the drill's shifting slightly in your hand.

The carburetor should be primed with gasoline to avoid having to crank the engine to fill the fuel line. If the engine has been put together correctly, it should start right up. If the engine refuses to start, check the ignition timing and other tune-up steps to make sure the engine is delivering a proper spark.

Engine Detailing

Besides painting the engine and accessories with the original color scheme, engine detailing consists of running the spark plug wires through the proper guides, putting new decals on the valve covers and air cleaner, routing the fuel and vacuum lines according to their proper path, and securing these lines with the correct clips installed at the original positions on the engine. You can get a guide to the location of the spark plug wires, fuel lines, and decals from the engine photos sometimes included in original sales brochures and by photographing the engines of properly restored trucks.

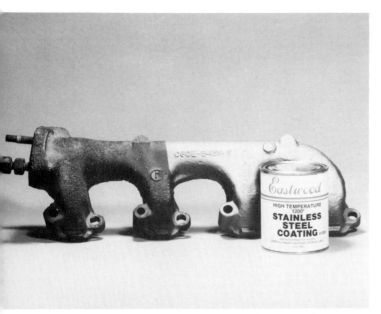

An engine's exhaust manifolds can be given a factory-new appearance with a stainless steel coating that applies easily and withstands high temperatures. The Eastwood Company

---

Good-looking Exhaust Manifolds

It's not uncommon to see a show car or truck with an engine painted to look as if it just came from the factory—and exhaust manifolds that look as if they've just been hauled out of a scrap yard. The problem with getting the exhaust manifolds to look as sharp as the rest of the engine is that most paints don't hold up well to the hot exhaust temperatures, and manifolds that have been cleaned by sandblasting soon rust.

The Eastwood Company markets a high-temperature Stainless Steel paint specially formulated for exhaust manifolds. Instructions with the paint advise sandblasting the manifolds to give the best surface preparation. The instructions also suggest cleaning the metal with a prep solvent before applying the paint. Although the paint can be diluted and sprayed, spreading it on with a foam applicator is recommended. This is the approach I used, and the paint went on smoothly and easily with an inexpensive ($0.39) applicator.

The stainless steel paint dries quickly, and once dry, the manifolds can be handled without leaving marks in the paint. The instructions call for baking parts 15 minutes at 400 degrees Fahrenheit to set the paint; a kitchen oven can be used—but leave a window or outside door open for ventilation, and warn the resident cook. The instructions also note, however, that exhaust manifolds or other high-temperature parts do not need to be taken through the baking step.

The painted manifolds have a fresh metal look that is very attractive. Best of all, the paint stands up to engine temperatures.

Chapter 8

Driveline Overhaul

Rebuilding an automatic or manual transmission or the rear end of a light-duty truck requires skills and specialized tools that few hobbyists possess. This means that if your truck shows wear in the driveline components, you will need to have them rebuilt by a professional. But even though you may not be rebuilding these components, you will still need to inspect them to determine whether or not the transmission and rear end need overhauling or can be used as is.

The first part of this chapter describes noises and other signs of transmission and rear end problems. It also gives guidelines that may save you money and headaches in getting these components rebuilt. In addition, it provides tips for removing the transmission and driveline from the truck. For actual step-by-step removal instructions, you'll want to refer to the Ford shop manual for your year and model truck.

Later in this chapter, you will find a discussion of the benefits of overdrive transmissions and approaches to installing an overdrive transmission in your vintage Ford truck. Overdrive, which was a factory accessory during the fifties, gives greater fuel economy, reduces engine wear, and offers higher highway cruising speeds. It's a driveline option well worth looking into for your truck.

Automatic Transmission Problems

By the nature of their design, automatic transmissions—first available on Ford trucks in 1953—are subject to different kinds of wear and failure than manual transmissions. If your truck is equipped with an automatic, three quick checks for trouble are to look underneath for signs that the transmission is leaking fluid, notice whether the truck slips in gear or slams into gear, and inspect the color of the transmission fluid.

Leaking fluid can usually be seen in puddles on the ground or garage floor or in traces of dripping fluid at the end of the transmission case or front of the drive shaft (leakage from the rear of the transmission is more likely to occur when the truck is parked on an incline). Leakage around the pan (at

If the truck you are restoring has worked a long, hard life, plan on a transmission rebuild or look for a low-mileage replacement unit.

Manual transmissions are not particularly complex. With the help of instructions from the Ford service manual, you may decide to rebuild your truck's transmission yourself.

Replacement bearings should be available through a local auto parts store. If you bring in the old bearing, the parts clerk can take the measurements and look up a replacement on an interchange chart.

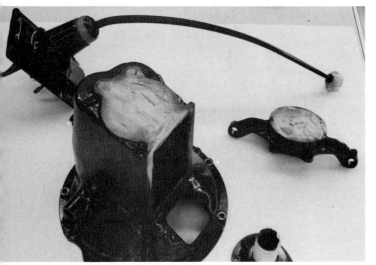

You will want to clean and refinish driveline parts before replacing them in your truck. During refinishing, use masking tape to cover mating surfaces, as shown here.

the bottom of the transmission) or the rear seal is relatively easy to fix with the transmission left in the truck. Slippage and hard shifting are signs that the transmission needs to be rebuilt. The fluid level of an automatic transmission is checked with the engine warmed up and running. With the truck parked on level ground, the fluid should reach the full mark on the dipstick. The fluid's color should be dark red. A brownish color—typically accompanied by a burned smell—means that the transmission is slipping. Burned, brownish-looking fluid is a sign that the automatic transmission is in need of a rebuild.

Manual Transmission Problems

Wear in manual transmissions is largely identified by noises. If you hear a rumbling sound coming from the general area of the transmission when the engine is idling with the clutch out, the front transmission bearing is worn and needs replacing. A similar sound that is heard from the same location when the truck is being driven in third gear—and that isn't heard with the transmission in neutral—indicates a worn rear transmission bearing. A chipped or missing tooth on first or second gear can often be heard as gear noise when the truck is operated in either of these gears—although you can expect to hear gear noise from any high-mileage transmission.

Besides worn bearings and worn or chipped gears, manual transmissions can also develop excessive end play (movement of the gears and shafts inside the case) owing to normal wear of the thrust washers and synchronizer clutches—or possibly caused by broken snap rings that are used to hold the bearings in place. End play is especially noticeable when it causes the transmission to jump out of gear. In this situation, you are driving along, let up on the gas momentarily, and the transmission "shifts itself" into neutral.

Although the problems described above can be caused by hard driving or by operating the transmission when it is low on gear oil, they are also the natural results of wear and high mileage. If the truck you are restoring has worked a long, hard life, plan on a transmission rebuild or look for a low-mileage replacement unit.

Transmission Disassembly

If you are disassembling your truck for restoration, the transmission typically comes out of the chassis after the engine. If you are rebuilding the truck and the engine doesn't need to be pulled, but the transmission shows signs of needing rebuilding, on 1942 and later Ford trucks, you can remove the transmission without disturbing the engine by uncoupling the drive shaft and then unbolting the transmission from the bell housing and dropping it down from underneath. You can remove the transmission this way on 1941 and earlier Ford trucks as well, but to get at the transmission, you will need to loosen the rear spring bolts and slide back the torque tube drive and rear end.

Transmission Parts and Service

When having a transmission rebuilt, look for a shop or repairer with experience on transmissions of your truck's vintage. Gasket sets and bearings for Ford manual transmissions are readily available from specialty suppliers, and you may even be able to purchase these parts at your local auto supply store. New gear sets may be expensive and hard to find. For this reason, it's usually less expensive to replace a high-mileage manual transmission with a lower-mileage unit. For later-model trucks with a Ford-O-Matic transmission, truck and car units interchange, making parts quite plentiful.

Clutch and Pressure Plate Servicing

The clutch makes a coupling between the engine and the driveline. When the clutch is engaged (the pedal is released), it causes power from the engine to flow through the transmission to the rear end. When the clutch is disengaged (the pedal is depressed), the engine turns freely from the driveline.

The clutch consists of a round plate containing friction material. When the clutch pedal is released, this friction material is forced tightly against the flywheel. The clutch then turns with the engine, causing the transmission input shaft, which fits through a splined collar in the center of the clutch, to turn also. In this way, engine power passes to the transmission.

Basically clutch problems are of three types. The first is clutch wear, which is felt as slippage. The second is clutch chatter, which is felt in an annoying jumping motion as the clutch is engaged. The third is wearing of the clutch mechanism rather than the clutch itself, which is often heard as a high-pitch squeal.

Clutch wear and slippage are easy to recognize. When the clutch slips, the engine turns faster than the driveline. If a truck is driven with a slipping clutch, you will notice a burning smell caused by the friction surfaces on the clutch sliding against the flywheel. Slippage is a sure sign that the clutch needs to be replaced.

Chatter can have several causes. The most common is an oil leak at the rear main bearing seal. When this seal leaks, the clutch facing often picks up oil, which causes the friction material to "jump" before seating tightly against the flywheel. Chatter can also occur if the clutch's friction material has developed hard spots. A third cause of chatter is a flywheel with a warped friction surface or a flywheel that isn't seated squarely on the end of the crankshaft.

Wear in the clutch mechanism can be heard as a high-pitch squeal or chatter when the clutch is disengaged. This noise is coming from a dry or worn pilot bushing or throw-out bearing.

If your truck exhibits any of these clutch problems, you'll need to remove either the engine or the transmission to get at the clutch mechanism. When servicing a worn clutch, it's a good idea as well to remove the flywheel and have the friction surface checked for trueness at a machine shop. The pressure plate should also be inspected for worn "fingers" (these are the prongs that make contact with the throw-out bearing when the clutch is disengaged), grooves and other signs of wear on the contact surface, or warped or broken springs (these are a sign of severe clutch abuse). The pilot bushing should in addition be checked for wear and replaced if necessary. Be sure to lubricate the pilot bushing before installing a new clutch.

Replacing a clutch is within most hobbyist mechanics' skills. The only specialized tools required are a torque wrench and an alignment shaft to center the clutch disc on the pressure plate. The service manual for your year Ford truck gives a step-by-step guide to follow if you're replacing a clutch for the first time.

Universal Joint and Drive Shaft Servicing

Universal joints are included in the driveline to change the angle of the power flow from the transmission to the rear end. When kept properly lubricated, universal joints will last nearly indefinitely. When lubrication is ignored, these constantly moving parts can wear rapidly.

Usually the first sign of universal joint wear is vibration in the driveline. This occurs because the needle bearings in the universal joints have worn enough to throw the drive shaft slightly off center. When the universal joints wear to the point where they are making noise, the bearings are gone and you're not going to be driving your truck very many miles before the universal joints fail and the drive shaft drops onto the road.

It's possible to replace the universal joints and still have driveline vibration. The cause in this case would be an out-of-balance, bent, or improperly installed drive shaft. You can have the drive shaft balanced at a shop specializing in the work, typically one that modifies heavy duty trucks. When installing the drive shaft in the truck, the sliding yoke goes at the upper end, behind the transmission.

Differential Servicing

Like that in other drivetrain components, wear in the differential—also called the rear end or third member—is also signaled by noises. Basically three kinds of noises indicate rear end problems. A howling sound—not a scream, but a noise that sounds like poorly meshing gears—indicates gear wear in the differential; this noise will normally change as you accelerate and decelerate, thereby altering the load on the ring gear and drive pinion. A rumble or rougher sound indicates worn bearings. A *clunk*, heard when you start to back the truck up, is a sign of backlash or wear that may be occurring anywhere in the drivetrain but is often located in the rear end.

The other problem to look for is oil leakage. This may be seen around the differential "pumpkin" (the gear housing in the center of the rear axle), where the cause is either loose bolts around the differential cover or a bad gasket. Drive pinion seal leakage will sometimes cause oil to be thrown from the rear universal joint yoke when the truck is running. This can cause the oil level to drop below the drive pinion in the differential. Oil can also leak past the seals at the ends of the axles. When this happens, the leakage is sometimes seen on the rear wheel backing plates or can be seen when the rear brake drums are pulled. Along with looking for signs of oil leakage, check the oil level in the differential. If the oil level is low, damage can occur to the rear end bearings and gears.

Gear noise can often be quieted or eliminated by adjusting the mesh between the ring and drive pinion gears. The rough use that pickups sometimes receive causes differential wear that is not normally found in cars. Sometimes trucks are even operated with different-sized tires on the rear wheels. If this has happened to your truck, the differential pinion and side gears are likely to show more wear than would be expected for the truck's mileage. This is a common cause of backlash or the *clunk* sound mentioned above.

When listening for bearing noise, don't get confused with normal tire noise. If your truck is fitted with mud and snow tires or if you are driving on rough pavement, the whine you hear is probably from the tires. To avoid confusing noise coming from bearings inside the differential with noise coming from the rear wheel bearings, turn the truck sharply left, then right on smooth pavement. Wheel bearing noise gets louder as the truck is turning. If you hear a constant bearing or gear noise that sounds as if it is originating in the rear end, the differential should be disassembled for inspection and repair.

One easy way to tell if looseness or backlash is coming from the differential is to jack up the rear end and have a friend put the truck in gear and slowly engage the clutch while you watch how far the drive shaft turns before the wheels begin to rotate. Unless you want your truck to be in factory-new condition, some backlash is tolerable, as long as the rear end doesn't howl.

It's possible to overhaul the rear end without taking it out of your truck—but if you are doing a frame-up restoration, you will probably remove the rear end anyway to clean it and to sandblast the frame. Rear end repair, like transmission overhaul, is work best left to professionals. Since the basic design of rear ends for light-duty trucks has changed little

A leaking drive pinion seal will sometimes cause oil to be thrown from the rear universal joint yoke when the truck is running. This oil may collect on the underside of the bed and is usually seen as a grease coating on the differential. Evidence of an oil leak means that the pinion seal for the differential drive needs to be replaced.

over the years, any repair shop with experience on Ford trucks should be able to perform this work satisfactorily.

Although gaskets and bearings should be readily available, you may have more difficulty locating replacement gears. In this case, you'll probably substitute a rear end from a lower-mileage truck.

Benefits of Overdrive

From the twenties, when Ford invented the pickup, until the seventies, pickups were most commonly found working on farms and at construction sites and doing service and delivery duty for a variety of businesses from appliance stores to lumberyards. It has only been since the early seventies or so that Americans have purchased pickups in large numbers as recreation and personal-use vehicles.

Because they were bought to work, the pickups we now restore and enjoy as collector vehicles are plagued by one shortcoming: their low rear end gearing. To the farmer who used a pickup to drive into the fields, low gearing was a plus. It meant less likelihood of spinning the rear wheels when starting out and therefore less chance of getting stuck in soft ground. Low gearing also better fitted the power curve of the older, slower-revving engines, thereby making these light trucks better able to handle a load. When the owner drove her or his pickup to town—a distance of maybe 10mi—the low gearing kept comfortable cruising to 40mph or 45mph. In those days the highway was only a two-lane road and probably had other slow-moving traffic—and besides, who was in a hurry?

Today that low rear end gearing means either a howling engine if attempts are made to push the truck to highway speed limits, or, if engine life is important, puttering along as though you were on a Sunday afternoon drive. Ideas that some pickup owners have about swapping in a car rear axle to gain higher cruising speeds usually don't work. So what's to be done? Pickups are fun vehicles. How can we get them to perform like a passenger car? After all, that's really how we use pickups today.

Fortunately that low gearing headache has a cure. It's the same cure that many Ford car owners turned to in the forties and fifties when they were looking for faster road speeds and longer engine life. The solution I'm talking about is an overdrive transmission.

What's an overdrive? Basically this is an auxiliary transmission that steps up the output speed of the engine—by 22 to 33 percent. This means that if your pickup—or car—has a comfortable cruising speed of, let's say, 45mph to 50mph, the addition of an overdrive transmission will boost that cruising speed to 60mph to 65mph with no increase in engine work. Sounds like magic, and in a way, it is.

Overdrive is also a way to better fuel economy, and that's another reason for installing the auxiliary transmission. Like road speed, fuel economy wasn't a big priority of original light-truck buyers, who looked mainly for reliability and ruggedness. It's not unusual for older light trucks to gulp fuel at the rate of 14 to 16mpg—and worse. Owners of pickups built in the late sixties and early seventies, when federally mandated emission controls first went into effect, often report gasoline guzzling figures of 8 to 12mpg. Given today's fuel prices and concern for conserving these resources, gasoline consumption in this range puts a damper on cruising and extended pleasure driving. Although overdrive doesn't work a miraculous "fix" on the engine, by cutting engine rpm to maintain highway cruising speeds, adding an overdrive transmission does increase fuel mileage. As the conservation billboards say, slower speeds mean more miles per gallon. With an overdrive

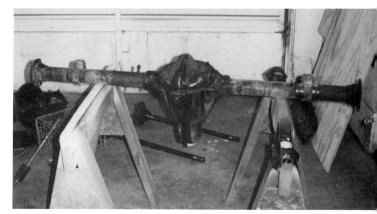

While the truck is apart, it's a good idea to pull the axles and check the condition of the bearings and seals.

The rear end assembly should also be cleaned and repainted before it is reinstalled in your truck.

transmission, the vehicle doesn't need to go slower, because the engine does.

The way an overdrive works is ingenious. It uses the output from the primary transmission to spin a planetary gear set that turns inside a larger ring gear. The result is a step-up gear ratio that sets the output shaft from the overdrive transmission turning significantly faster than engine speed. At the rear end, this faster drive shaft rotation translates into higher road speeds.

Although Ford cars could be fitted with a form of overdrive in the thirties, this higher gearing alternative didn't become a factory option until Ford switched to the Hotchkiss-style, or open, driveline in 1949. Overdrive didn't appear as an option in Ford light trucks until the mid-fifties.

When installed, the overdrive transmission sits behind the standard three-speed and is activated by a combination electrical and mechanical control system. Like other domestic manufacturers, Ford purchased its overdrive units from Borg-Warner.

Because car and light-truck engine and transmission combinations are similar, overdrive can be retrofitted to a Bonus Built series or later pickup without a great deal of difficulty. If a truck overdrive isn't available, the same unit from a car will work. But you can't just bolt an overdrive gearbox to the back of your truck's three-speed. The two transmissions were designed as a pair, with the overdrive's companion three-speed gearbox having a longer output shaft that turns the overdrive gear cluster. If you're thinking of retrofitting overdrive to your Ford F-1 or F-100 truck, the place to start is to locate a three-speed and overdrive transmission. Scrap yards are a good source, as are swap meets and ads in old-car and -truck hobby magazines.

Retrofitting overdrive to trucks where this transmission option was available originally is relatively simple, provided you are converting a standard three-speed to three-speed and overdrive. The standard three-speed will come out by removing the drive shaft at the universal joints, disconnecting the speedometer cable and shift linkage, and unbolting the transmission at the bell housing. Installing the overdrive unit is a simple matter of reversing these steps.

Note that some overdrive installations require an extra cross-member to support the longer transmission. If this is the case for your truck, you will also need to fabricate the cross-member.

Unless you also locate a drive shaft from an overdrive-equipped truck of the same model, you'll have to shorten the drive shaft and the shortened shaft must be balanced. You will need to have the drive shaft shortened and balanced by a shop specializing in driveline modifications. A control cable also needs to be run from the dash to the overdrive lever to engage and disengage the auxiliary transmission. It shouldn't be too difficult to find an overdrive lever—stamped OD—and cable in a scrap yard, or you can buy a suitable cable control at an auto parts store.

If locating and installing an overdrive transmission isn't a problem, what keeps every light-truck owner from retrofitting this cure for low rear end gearing and opportunity for better fuel economy? The problem comes in finding the electromechanical control pieces. These Borg-Warner–designed overdrive transmissions used a rather complicated combination of solenoid, relay, and throttle-operated kickdown switch to engage and disengage the overdrive, and locating all of

Although Ford cars could be fitted with a form of overdrive in the thirties, this higher gearing didn't become a factory option until Ford switched to the Hotchkiss-style, or open, driveline in 1949. The overdrive transmission shown here is installed in a 1953 Ford F-100.

The control device for a factory-installed Borg-Warner overdrive is the electric solenoid pictured here. Replacement 6-volt solenoids are not being made, so if you find an overdrive with a missing or burned-out solenoid, you can expect to pay a premium—if you can even locate a working solenoid.

these control pieces in working condition is much harder than turning up the overdrive transmission. The good news, however, is that overdrive can be installed easily without using the electrical controls and kickdown linkage.

The way around the original electromechanical control setup is to use a completely manual control device available from Overdrives. This manual control unit eliminates all the electrics as well as the kickdown switch. It also gives double the number of gear ratios, so that instead of offering three forward speeds plus overdrive, your truck can now be driven as though it had a six-speed transmission. You won't shift through all six potential gears each time you

How to Drive in Overdrive

In addition to ensuring cruising speeds in the range of today's traffic, greater fuel mileage, and longer engine life, an overdrive transmission can also heighten the enjoyment of driving your truck. To engage the overdrive, just shift through the gears as you normally would, and once in "high," accelerate to 30mph or so for a Borg-Warner overdrive, or 47mph for a Gear Vendors underdrive-overdrive, and let off the gas. When you resume acceleration, you'll notice that the engine revs have dropped as though you had shifted into a higher gear.

With an original equipment of manufacturer (OEM) Borg-Warner overdrive, whenever the auxiliary transmission is engaged, you also have freewheeling—which means that the truck coasts whenever you let up on the gas. One chief advantage of freewheeling is a smoothness that makes the manual transmission feel like an automatic. If you lift your foot from the accelerator pedal, the truck won't "buck" and slow down; it'll just gently drop off speed. Also, if you get the engine revs right and keep a load off the gearbox, freewheeling will allow you to shift from first to second and from second to third without using the clutch—so in this sense, it is very much like having an automatic transmission.

Of course, the disadvantage is that when freewheeling is operating—and this is any time the overdrive transmission is engaged, or the overdrive lever is pulled out—engine compression doesn't help the truck brake. This means that if you are driving in hilly or mountainous terrain, you'd better keep the overdrive locked out because on a long downgrade without the engine helping hold the truck back, you could burn up the brakes. For around-town or distance driving on fairly level ground, the brakes should be able to handle whatever slowing or stopping needs you encounter, without the engine's help. If you are operating an overdrive-equipped truck, you'll need to remember either to disengage overdrive or to apply the emergency brake when parking. Forgetting to do one of these may mean that when you come back for the truck, you'll find that it has "moved itself" to another spot. With the overdrive lever pushed in, no engine compression is available to keep the truck from rolling—which it's likely to do if it is parked on a surface that's not level.

One other word of caution: Overdrive must never be engaged when shifting into reverse. This is crucial. If the overdrive is engaged and reverse is attempted, the auxiliary transmission can bind up or destroy the overdrive mechanism. This concern only applies to a manually controlled Borg-Warner overdrive. The electrical controls, as set up by the factory, automatically disengage overdrive when the vehicle comes to a stop. There's also another reason to make sure overdrive is disengaged when stopping: the auxiliary transmission is not designed to pull a static load.

If you set up the overdrive as a gear splitter—a feature that a manual control brings to a Borg-Warner overdrive and a built-in feature of the modern Gear Vendors underdrive-overdrive transmission—you'll find that your truck can be driven just like a larger truck that's equipped with a two-speed rear axle. This means that for every regular, or underdrive, gear, you also have an overdrive. For example, when motoring around town, you may find second overdrive to be an ideal gear for fuel economy and power; to reach second-over, just shift from first to second, then let up on the gas to engage overdrive.

If you want to play "big trucker," you can have fun with your overdrive's gear-splitting feature. Sometimes, after engaging second-over, I shift to third and disengage the overdrive (push in the control). Then as I leave the city limits and move out onto the open road, I pull out the manual control and shift into third-over, the cruising gear. Sort of a Walter Mitty feeling comes from playing the overdrive and three-speed like a road hauler moving through the gears. "Better tell 'em I'm coming by," I think as I reach for the air horn lanyard.

If you think an overdrive transmission has appeal, you'll probably base your decision of whether or not to go to the expense and trouble of installing the auxiliary transmission in your collector truck on how you plan to use the truck. If you expect to take your truck out on the interstate and travel 100 or more miles at a stretch to attend a show or just to enjoy turning back the clock, then you should seriously consider adding an overdrive transmission. The reduction in engine noise and cruising effort, plus increase in fuel economy that you will experience will probably make you wonder why every truck wasn't fitted with overdrive at the factory.

pull away from a light (as explained in the "How to Drive in Overdrive" section later in this chapter), but you do have twice as many gear ratios available for special purposes.

Retrofitted Overdrive

Through 1941, Ford used a torque tube driveline that, because it is incompatible with a normal overdrive setup, makes installation of the auxiliary transmission more difficult. (Torque tube drive means that the drive shaft is enclosed in a tubular housing that forms a rigid support member for the rear axle.) But if you have a thirties or 1940–41 Ford pickup, there is still a way for you to install an overdrive transmission.

A small specialty shop called Overdrives fits Borg-Warner overdrives to practically any vehicle with four wheels. The operations performed by this company differ in two important ways from the factory approach. First, the overdrive is installed in front of the differential, rather than behind the transmission. Second, Overdrives replaces the electrical solenoid with a manual unit. The reason for the manual control is the lack of a supply of dependable solenoids; no one is rebuilding them, and the electrical control is missing from about 90 percent of available overdrive transmissions.

To have Overdrives install an overdrive transmission in your Ford pickup's torque tube driveline, you will need to send the torque tube, plus the drive shaft and rear end carrier. If you also provide the overdrive, you will receive a rebate for that unit. When your parts and order are received, the torque tube is cleaned, sandblasted, and painted. The overdrive to be mated with the torque tube is likewise cleaned, disassembled, and rebuilt. This done, a flange matching the diameter and hole pattern of the differential is welded to the rear of the overdrive housing. Then the torque tube is shortened the distance of the overdrive unit. The drive shaft is also cut this same amount, and a sprocket is welded to the end that will turn the overdrive. A matching sprocket is welded to the overdrive input shaft. A double chain connects the two sprockets and allows for a small amount of play to compensate for any misalignment in the driveline. Finally everything is assembled and the flange at the end of the overdrive housing is machined as needed to make sure the overall length is the same as that of the original torque tube.

Retrofitted Overdrive Installation

To install the overdrive setup in your truck, you will reverse the procedure you followed to remove the torque tube and rear end carrier.

Once the overdrive is installed, making it functional is a simple process of running two cables from the dash to the overdrive controls. One of these cables is supplied with the overdrive setup. You will need to buy the other, from either an auto parts store or a scrap yard. You will need to decide whether you want to mount the cable controls on the dash panel or on a bracket attached under the dash. Mounting the cable controls in the dash requires

The heart of an overdrive solenoid is the electric coil—partially unwound here to show the amount of wire it contains. When a solenoid fails, the reason usually is a break in the insulation or the wire someplace in this coil.

This simple manual control unit, available from Overdrives, replaces the electric solenoid, wiring, relay—all the potentially troublesome electronics.

drilling extra holes. If you know of a scrap yard with forties and fifties cars and trucks, you should have little difficulty finding brackets to hold the overdrive controls—and you should be able to buy the overdrive control cable while you're at it.

Both cables will need to be routed through the firewall, either through existing holes that might be in their path or by drilling new holes. Be careful to run the cables so that they don't interfere with the clutch or brake linkage and are free of kinks or sharp bends. When attaching the cables to the transmission controls, make sure the cable moves enough to engage and disengage the controls fully. The cable supplied by Overdrives operates the manual control unit. The other cable engages and disengages the overdrive's freewheeling feature.

When everything is hooked up, fill the overdrive transmission with gear oil before operating the truck.

Overdrive's Freewheeling Feature

A manufacturer-installed or retrofitted Borg-Warner overdrive transmission has a freewheeling feature that lets the truck coast whenever you let off on the gas. The advantages of freewheeling are greater fuel economy and the opportunity for clutchless shifting from second to third gear. The disadvantages are faster brake wear and the possibility of a runaway if overdrive is left disengaged in mountainous travel.

The freewheeling clutch is located at the rear of the overdrive gear cluster and consists of a set of twelve clutch rollers that press against an outer ring gear while the main shaft is receiving torque from the engine. When the engine torque cuts back (when you let up on the accelerator) the clutch rollers disengage and cause the power link between the main shaft and the output shaft to be disrupted. This allows the output shaft to turn with the differential without any braking restraint from the engine.

Troubleshooting of an Electrically Controlled Overdrive

Overdrive offers the same benefits to owners of collector trucks as it did to these trucks' original buyers. The problem a collector sometimes has with his overdrive-equipped vehicle is troubleshooting and correcting nonfunctioning units. Only rarely does the cause of a nonfunctioning overdrive lie with the transmission itself. The mechanism is simple and rugged and won't let you down on the road. A friend lost the overdrive in his car on a cross-country trip (the auxiliary transmission had run low on oil) and still drove the remaining 2,000mi to his destination in conventional drive. The likely cause of overdrive problems is with the electrical control circuitry. The following procedure outlines a sequence that can be followed to find and correct electrical problems with a Borg-Warner overdrive.

A Borg-Warner overdrive utilizes three electrical circuits in operating its controls. One circuit incorporates a governor and allows the overdrive to kick in above 25mph. Another circuit energizes the solenoid to engage the overdrive. A third circuit, activated by the kickdown switch, disengages overdrive. To check for problems in the electrical circuitry and controls, you will need:

- test light or VOM meter
- extra 30amp fuses
- a 3-4ft length of 12- or 14-gauge test cable with alligator clips soldered onto the ends
- a shop manual for your truck
- optionally, a *Chilton's* or *Motor* manual of your truck's vintage that covers overdrives
- patience and perseverance

With electrical problems, always check the simplest cause first. The relay has a fuse. Be sure the fuse is good; looks don't always tell. If the fuse is blown, replace it and try the overdrive. If the overdrive still doesn't engage, you can proceed methodically to troubleshoot the three control circuits.

To check the governor circuit, turn the ignition switch off and on. If the relay or solenoid clicks when the switch is turned on, there's a short in the circuit or the relay is defective. To check which is the problem, remove the wire from the relay to the kickdown switch (a wiring diagram will help here) and again turn the ignition switch. If the relay clicks, it is defective; if it doesn't, there is a short in the governor circuit.

To determine whether the solenoid is working, apply current to the "hot" terminal while grounding the case of the solenoid. This can be done under the truck or you can remove the solenoid and make the check at the battery. If the solenoid operates, it

If your Ford pickup is a pre-1942 model with a torque tube driveline, it's still possible to install an overdrive transmission. The setup shown here integrates the overdrive gearbox into the torque tube.

The trickiest part of mating a Borg-Warner overdrive to a Ford torque tube drive is welding the mounting flange and support bracing onto a 40- to 50-year-old casting.

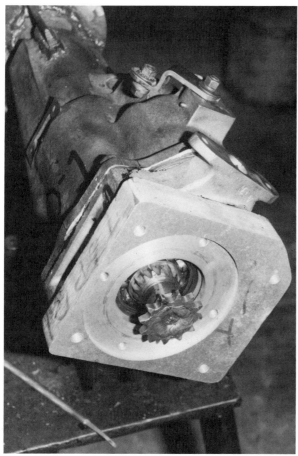

To couple the overdrive to the drive shaft inside the torque tube, a sprocket is welded onto the overdrive input shaft.

obviously is OK. If it doesn't, you need a new one. Replacements are no longer being made and are very difficult to find.

The easiest way to isolate overdrive electrical problems is to check for continuity (current flow) at each connection. If the solenoid is good, but doesn't operate, with the ignition switch on, use a test light or VOM to make sure current is passing to the relay. If the relay is "dead," there is a short in the wiring between the ignition switch and the ignition terminal on the relay.

As noted before, make sure the relay fuse is good. You can check the fuse with a test light or VOM by connecting the test device to the end of the fuse opposite the ignition lead and checking for continuity. When making a current check, be sure to touch the other lead of the test device to a good ground.

To test the relay, connect the test device at the solenoid terminal and ground and use the wire with alligator clips to ground the kickdown terminal. If the test device does not indicate current flow, the relay is defective.

Now check for continuity at the solenoid by grounding the kickdown switch terminal at the relay. If the relay is working properly, the solenoid should click when the kickdown terminal is grounded. If it doesn't, and the solenoid is good (as indicated by an earlier test), there is a break in the wiring between the relay and the solenoid.

The kickdown circuit grounds engine ignition momentarily when the accelerator is pressed to the floor. The brief power interruption allows the solenoid to disengage the overdrive. To check the kickdown circuit, first observe whether the wire running from the kickdown switch to the coil is connected to the DIST side of the coil. If the wire is

A matching sprocket is welded onto the drive shaft. A double chain connects the two sprockets, allowing for a small amount of play to compensate for any misalignment in the driveline.

missing or connected to the BATT side of the coil, properly connect the wire. Next, remove the wire from the kickdown switch at the solenoid (refer to the service manual wiring diagram for the wire's color code). While the engine is running at fast idle, reach under the accelerator pedal and press the overdrive switch until it bottoms. The engine should stop. If it doesn't, there is a short in the wire from the distributor to the kickdown switch, or between the switch and the solenoid.

Assuming you have checked everything and the electricals are OK, but there is still no OD, the problem has to be mechanical. An overdrive is quite a lot simpler mechanically than a manual three-speed transmission. If you've done the mechanical work described so far in this book there's no reason you couldn't overhaul an overdrive transmission. Detailed instructions are found in a Ford truck or car shop manual for years in which overdrive was offered as an option (mainly the 1950s) or in a *Motor* or *Chilton's* manual of that vintage.

Modern Overdrive Installation

Ford trucks of sixties vintage are becoming popular with collectors, and these more modern trucks can also benefit from the advantages of an overdrive transmission—particularly in fuel economy. For these newer trucks, rather than retrofit an old Borg-Warner unit, you have the option of installing a newly manufactured overdrive.

Although several aftermarket overdrives are on the market, most have limited applications for older vehicles. The most versatile—and most ruggedly engineered—unit is an underdrive-overdrive transmission from Gear Vendors, which is a popular aftermarket product for recreational vehicles. This auxiliary step-up transmission can be coupled to the venerable Ford C-6 automatic and various manual transmissions for trucks going back to the sixties. The Gear Vendors transmission also can be—and has been—fitted to earlier vehicles. In fact, many of the cars and trucks entering the Great American Race are able to travel cross-country at interstate highway speeds because of the installation of a Gear Vendors underdrive-overdrive transmission.

Among its many attractions, the Gear Vendors underdrive-overdrive can be coupled to automatic as well as manual transmissions. It also works with four-wheel drive. Like original equipment overdrives, the Gear Vendors unit attaches to the back of the transmission, but unlike the factory-installed overdrives, the Gear Vendors product simply bolts to the primary transmission's output shaft through a special adapter that replaces the manual three-speed, manual four-speed, or automatic transmission's tail shaft housing.

Its being a completely separate transmission means that the underdrive-overdrive is a sealed unit that has its own lubrication and control systems. The separate unit feature is important. Some modern overdrives share lubrication with the primary transmission. This means that if either transmission fails (the clutch packs or bearings grind up), contaminated lubricant circulates through both transmissions, potentially destroying both. Should a Gear Vendors underdrive-overdrive transmission fail mechanically—a highly unlikely condition given the transmission's extremely rugged and reliable design—the vehicle can still be driven, though without the overdrive advantage.

A Gear Vendors underdrive-overdrive transmission can be ordered directly from the manufacturer or from a number of dealers located around the United States. Dealers can install the transmission in your truck, or if you are somewhat mechanically adventuresome, you can install the auxiliary transmission yourself.

When ordering the underdrive-overdrive from the manufacturer, you will need to know the type of transmission that's in your truck, and the rear end ratio. Specs on the transmission should be listed in the service manual for your truck, and the rear end ratio is often found either on a tag bolted to the differential or stamped in one of the axle housings.

If you decide to install the Gear Vendors underdrive-overdrive transmission yourself, you'll receive from Gear Vendors or its dealer the underdrive-overdrive transmission, the adapter for your truck's transmission, an extension for the speedometer cable, a new speedometer gear

The Borg-Warner overdrive transmission has a freewheeling feature that increases gasoline mileage and allows clutchless shifting from second to third gear. The freewheeling clutch can be seen here in front of the overdrive gear cluster.

calibrated to your truck's rear end ratio, a speed sensor that will be coupled between the end of the existing speedometer cable and the new extension, an electronic control box, and a foot switch for shifting in and out of overdrive, plus a couple of Gear Vendors stick-on nameplates in case you want to advertise the underdrive-overdrive feature on your truck.

Begin the installation by parking the truck in the work area, disconnecting the battery, raising the truck to a height that allows you to work comfortably underneath, and placing jack stands under the axles to support the truck at this elevation. It won't be necessary to raise the vehicle very much; a couple inches' clearance between the tires and the shop floor should give you enough room to move freely underneath the truck. Be sure to support the truck on professional-quality jack stands, and never work under a vehicle that is elevated by stacks of wood, concrete blocks, or some other makeshift arrangement that can collapse and pin you under the vehicle.

The Gear Vendors installation manual goes into quite a bit of detail on measuring the drive shaft angle to make sure the additional length of the auxiliary transmission doesn't pitch the drive shaft at an angle severe enough to cause universal joint stress or driveline vibration. It also gives instructions for measuring the length of the existing drive shaft to make sure the shortened shaft will be within acceptable limits. You will want to study these instructions and "engineer" the installation before proceeding. If your truck has a two-piece drive shaft,

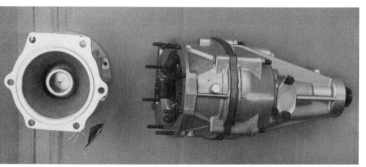

Newer Ford trucks can also benefit from the advantages of an overdrive transmission through the installation of a modern aftermarket unit made by Gear Vendors.

The Gear Vendors overdrive bolts to the back of the truck's manual or automatic transmission. The installation is straightforward, and the electronic controls hook up through simple phone jack connectors.

Installing an overdrive transmission requires having the drive shaft shortened by the length of the auxiliary transmission.

it may be necessary to eliminate one of the shafts, going to a one-piece drive shaft with no center bearing. In most cases, the only modification required is having the drive shaft shortened by approximately 14in—the length of the Gear Vendors transmission.

When you have worked through the logistics, you will start the actual installation procedure by removing the drive shaft, which is accomplished by loosening the universal joint yokes at the rear axle. This done, you can drop the drive shaft by pulling it out of the slip yoke at the transmission, and put it aside for the time being. Next you will disconnect the speedometer cable from the transmission, then you will remove the universal joint yoke and tail housing at the transmission. If your truck has an automatic transmission, you'll want to put a pan underneath the tail housing to catch any transmission fluid that may seep out as you loosen the tail housing bolts. You won't be draining the fluid from the transmission, and it's likely that only a small amount of fluid will have seeped into the tail shaft housing.

This is the extent of the disassembly steps. Now you're ready to install the underdrive-overdrive. After lubricating the coupler in the new tail shaft adapter, installing the underdrive-overdrive is a bolt-up process. First the adapter is bolted to the back of the primary transmission, in place of the tail shaft, and then the overdrive is bolted to the adapter. To keep the bolts from working loose, place a dab of Loctite Blue on the bolt threads.

It's all a very straightforward process. No complicated aligning is needed, and the Gear Vendors castings are machined to precise fits. A gasket is supplied to place between the adapter and the overdrive; you'll use the old gasket, or a replacement, between the transmission and the adapter.

With the overdrive transmission in place, careful measuring is required to determine how much the drive shaft needs to be shortened. With the proper dimensions, you can have the drive shaft shortened at a shop that performs this service. This is a more critical step in the overdrive's installation, and it's important to find a shop that has experience in this type of work and can also balance the shortened shaft. When you get the shortened drive shaft back from the driveline shop, install it in the reverse order from that in which it was removed.

The remaining steps are to hook up the electronic controls, which consist of a switch unit that mounts in a convenient location under the front edge of the dash, a "black box" that can be mounted on the firewall or kick panel near the fuse box, and a foot control switch that mounts on the floorboard. After topping up the underdrive-overdrive with Dextron II automatic transmission fluid, it's off for a test drive.

If the Gear Vendors underdrive-overdrive is coupled to an automatic transmission, it can be operated in two modes: automatic or manual. In automatic mode, which is set by flicking a rocker switch on the control panel, the vehicle will shift through the gears in normal fashion, then climb into overdrive automatically at about 47mph. When overdrive engages, the engine speed will drop by 500rpm to 700rpm—yet the vehicle will maintain the same road speed.

In manual mode, overdrive can also be engaged between any of the transmission's lower gears. This means that the three-speed automatic now becomes a six-speed transmission. The extra gear ratios between first and second and second and third are called "underdrives"—hence the transmission's name, underdrive-overdrive. With a manual transmission, the overdrive and underdrive gears are selected by depressing the clutch momentarily and pressing a hand switch.

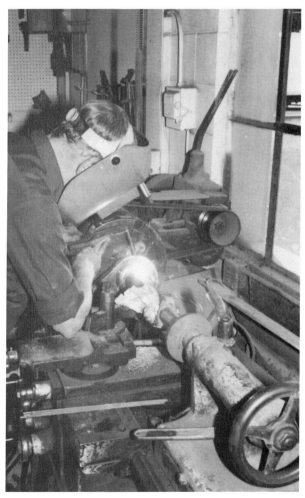

The job of shortening the drive shaft should be done by a driveline specialist who can not only make sure the shortened shaft is aligned correctly, but also balance the shaft to prevent driveline vibration.

Chapter 9

Rebuilding Brakes

An original brake system on a twenty-year-old, or older, truck is prone to failure—a very dangerous proposition.

Unless you are certain your truck's brakes have been *completely* overhauled, follow the procedure described in this chapter and give your truck a brake system rebuild. The other alternative is to hire someone else to do the work for you. Don't trust your safety, and that of your passengers, as well as your truck's preservation, to brakes that may be working marginally and could fail when you need them most.

It's not a good idea to take a patchwork approach to brake system repair. If one brake line springs a leak, probably other lines are at the point of springing leaks too. The recommended approach, and the one taken in this chapter, is to inspect and overhaul the entire braking system.

To understand what you'll be looking at, visualize what happens when you step on the brake pedal. As the foot pedal is depressed to apply the brakes, it presses a plunger in the master cylinder, which is mounted under the floorboard on Ford trucks with through-the-floor brake and clutch pedals and on the firewall of later trucks with firewall-mounted "swing" pedals. This plunger moves a piston to compress a small amount of brake fluid and in so doing, sends a pulse of hydraulic pressure through the brake lines to the four wheels. The brake lines terminate at the wheel cylinders. Here the fluid expansion forces the brake shoes out against the brake drums, slowing the vehicle. This is what happens in a properly functioning hydraulic brake system.

Now let's look at what can go wrong on a neglected, worn, or corroded hydraulic brake system. Starting again at the brake pedal, the pedal arm can have worn its shaft—from lack of lubrication—so that it binds when pressed and requires extra pedal pressure. This will make brake response slower. Then as the piston compresses the fluid inside the master cylinder, some of the fluid may leak past the piston seal—owing to corrosion on the cylinder walls. Since not as much fluid is compressed, the pulse of hydraulic pressure through the lines is reduced. Even with the reduced pressure, corroded brake lines could still spring a leak, causing the pressure in the brake system to normalize and preventing any braking action. Assuming the lines hold and the hydraulic pulse passes to the wheel cylinders, the wheel cylinder linings may also be corroded, so that pressure loss again occurs. If this happens, only weak braking action will result.

Worn linings or grooved, out-of-round brake drums will further deteriorate braking efficiency. The common condition is a combination of problems resulting in only marginal braking action—and almost certain brake system failure in the event of an emergency like the need to make a panic stop.

Brake Removal

Brake system overhaul typically starts at the wheels with removal of the brake drums. If you're not doing a frame-up restoration where every part will be stripped, cleaned, and refinished, and this is your first experience at a brake overhaul, it's a good idea to rework the brakes on one side of the vehicle at a time. This way, you can look at the other side for a guide to fitting all the parts back together correctly, if necessary.

Begin by loosening the lug nuts with the wheels on the ground, then jack up the truck and support it on jack stands. With the truck elevated on secure supports, remove the lug nuts and slip the wheels off the brake drums. To remove the brake drums, you first need to loosen the brake adjustment.

Brakes are adjusted in a couple of ways. The single anchor brakes found on half-ton trucks of the F-1 and F-100 series use star wheels to crank the shoes toward or away from the brake drums. The star wheel is located at the bottom of the shoes and is reached through a slot in the bottom of the backing plate. This hole may or may not be covered with a plug; often the plugs fall out or are not replaced after an earlier adjustment. Although a Z-shaped tool is made for turning the star wheels, a screwdriver can also be used.

The brakes on half-ton Fords through 1947 and

later heavier-duty Ford trucks with double anchor brakes adjust by turning two 5/8in nuts located on the upper right and upper left of the backing plates. Internally these nuts turn a cam that wedges the shoes toward or away from the brake drums. You'll feel when the shoes have been loosened by the free spinning motion of the drums.

Before removing the drums, you should put on a dust mask. The brake shoes commonly used on older trucks contain asbestos, and it's wise to take precautions against breathing the asbestos fibers that are found in the brake lining dust that you'll encounter when you've pulled off the drums. Since asbestos fibers can cause lung and breathing problems, you will also want to take steps to keep from stirring up the lining dust. This means that you will vacuum the dust out of the drums and from around the brake parts. Do not "sweep" out the dust with a paintbrush or whiskbroom. If brake lining dust falls onto the floor of your work area, vacuum it up, don't sweep it away. Sweeping stirs up the dust, increasing the chances that you or others will breathe in the asbestos fibers.

On the front, pop off the dust caps that cover the spindle nuts, if these covers are still in place. Mechanics usually remove these caps by grabbing them with channel lock pliers and flipping them loose, but this leaves the caps with dents that look like the creases in a highway patrol officer's hat. A better method is to insert a screwdriver between the cap lip and the brake drum hub and pry the cap loose by working the screwdriver around the hub.

With the cap out of the way, pull the cotter key from the spindle nut and turn the nut loose with an adjustable wrench, open-end wrench, or socket. Don't use pliers to loosen the spindle nut; their serrated edges will cut into the nut, making it difficult to fit the right tool onto the nut the next time you need to tighten or loosen it.

Once the spindle nut is removed, the drum should pull toward you easily. If it seems stuck on the brake shoes, and you have adjusted the shoes as loose as they will go, grab the edges of the drum and work it back and forth, over the shoes. The difficulty sometimes encountered is that the brake shoes can wear the drum so that a lip forms at the outer edge of

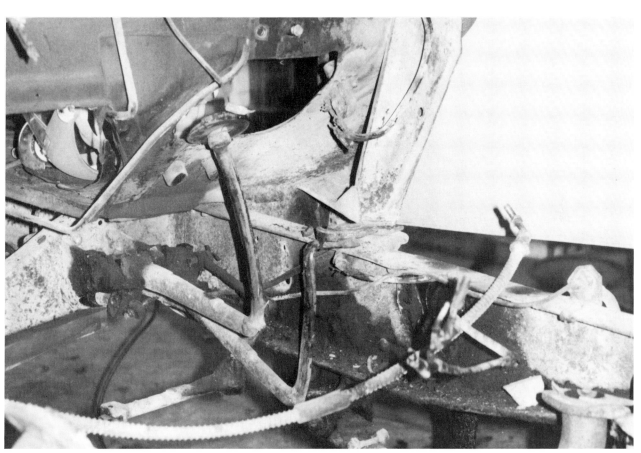

Often on an older truck, you will notice wear (in the form of side play) of the brake and clutch pedals on their shaft. This side play is not only annoying, it can cause the pedals to bind on the floorboards, creating a potentially dangerous condition. The wear is caused by a lack of lubrication of the pedal grease fittings over the years. The shaft area of the brake and clutch pedals on this truck shows a lot of caked-on grease—a good sign. If pedal and shaft wear is noticed, the pedals will need to be removed from the shaft and rebushed.

the brake sweep area. If the drums were not turned to cut away this lip when new linings were last installed, the lip can have become quite deep.

Where a lip has been allowed to form, and the brake adjustment star wheel or cams are rust bound and refuse to turn, the drums may catch on the shoes and refuse to pull loose. When this happens, the only way to get the drums off may be to cut the ends of the anchor pins that hold the shoes to the backing plate, with side-cutter pliers or a torch. Cutting the anchor pins will enable you to pull the drums loose, but the shoes will probably pop free of the wheel cylinder, which may also cause the wheel cylinder to come apart. This method for freeing the drums should be used only as a last resort.

As you slide a front drum off the spindle shaft, the large washer that sits behind the spindle nut and the outer wheel bearing will usually slide off the spindle. The washer and bearing should be picked off the spindle before they fall onto the shop floor, and placed in a container for safekeeping. It is important not to mix bearings from one front wheel to another, so a good idea is to place a container— clean, empty coffee cans or plastic milk jugs with their tops cut off work well—at each wheel to hold parts as they are removed. Now you can slide the drum free and place it on the floor in a nearby, but out-of-the-way, location.

Rear wheel drums on Ford light-duty trucks remove differently depending on the year of the truck. On light-duty trucks through 1947, they fit onto a tapered axle shaft and are held in place by a nut threaded onto the end of the axle. The recommended way to remove this style rear brake drum is with a wheel puller. This puller is a costly tool, however, and in the days before tool rental businesses became popular, small repair shops and "shade tree mechanics" used other methods to free the drum from the axle without a puller. One method was to loose the axle nuts and drive the truck in a series of tight circles until the wheels began to wobble. The other was to rap the ends of the axle with a heavy hammer. Some shade tree mechanics would rap against the loosened axle nut—often making it difficult to thread the nut back on—but more frequently, a "knocker" would be used. A knocker is essentially a length of steel rod that is bored and tapped at one end so that it can be threaded over the axle stub. Neither of these alternative methods is advisable.

Removing the axle nuts and driving the truck in circles can cause the drum to slide off the axle, creating considerable damage. Rapping the end of the axle with a hammer can damage the axle threads

A brake system overhaul typically begins with the removal of the brake drums. This is done by loosening the lug nuts, jacking up the truck, supporting the truck on jack stands, and then removing the wheels.

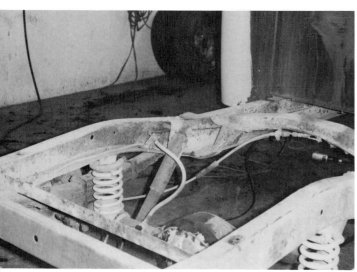

Chances are, you will replace all the brake lines. It is easy to inspect and replace the rear brake lines once the pickup box has been removed.

as well as the carrier bearings inside the differential. The right tool for this job is a wheel puller. It's best to use the type with arms that attach to the lug bolts. Using a puller that grips the brake drum by its outer edge risks warping the drum.

On light-duty Ford trucks from the F-1 series (1948) and later, the rear drums are held in place by three 5/16in bolts that thread into the axle flange. If the axle flange is grease coated—from leaking rear seals—you may not even see these nuts until you scrape off the grease. On heavier-duty three-quarter- and one-ton (F-250 and F-350) trucks, the brake drums pull loose after the lug nuts are removed.

As with the front, the rear brakes should also be loosened before attempting to remove the drums. On F-1 and later trucks, the drums sometimes rust to the axle flange and lug bolts. To break this rust bond, spray penetrating oil around the base of the lugs and around the hole in the center of the drum. Then tap the drum a couple of times around the flange area (the flat surface beside the lugs), and the drum should pull free. With the drums off, you can now see what's been happening when you pressed down on the stop pedal.

Brake Inspection

Typically what you'll see as you look at the brakes of an older truck are worn linings, often with just a thin layer of friction material showing; scored drums, sometimes with deep scratches in their contact surface; a great deal of lining dust coating the backing plates, brake shoes, and other internal surfaces; and possibly a thicker, darker coating—a mixture of brake fluid and lining dust—around the wheel cylinder. You may also see other problem signs such as broken return springs.

Brake Rebuild

Since the condition of your truck's brakes is crucial to the vehicle's safe operation, the overhaul procedure will be essentially the same whether you are restoring or rebuilding. The only difference may be in the thoroughness with which you refinish the brake parts that will be visible when everything is reassembled.

Fortunately brake parts for vintage Ford trucks are in good supply. You'll probably be able to purchase new brake springs, shoes, and wheel cylinders from a local auto supply store. If not, you're sure to find these parts through a vintage Ford truck parts supplier.

When rebuilding a brake system, the best policy is to replace all the operational parts. Before any new parts are installed, however, you will remove the brake mechanism and backing plate at each wheel, then clean and refinish the brake drums and backing plates. In a ground-up restoration, brake overhaul usually occurs as the last stage of redoing the chassis. With a rebuilding approach, the brakes should be overhauled as the first step to making the truck operational.

Brake Mechanism and Backing Plates Disassembly

Disassembling the brake mechanism at the wheels is a simple matter of removing the return springs, unhooking the clips that hold the shoes in place, and pulling off the shoes. The wheel cylinders attach to the backing plates with bolts that are reached from the back side of the plates. Once these bolts are removed, the brake line has to be disconnected from the wheel cylinder. At the front, the brake line connection is through a rubber flex hose that is usually hardened and cracked. If this is the condition of the flex hose on your truck, simply cut the hose. It can be removed from the wheel cylinder later. At the rear, the brake tubing connects directly to the wheel cylinder. If this brake tube fitting has become frozen to the wheel cylinder with rust, the line will have to be cut.

The backing plates are held in place by four bolts. Loosening these bolts allows you to pull the plates away from the spindle assembly. The next step is to clean the backing plates and prepare them for refinishing. Then you will clean and inspect the brake drums.

Wheel Bearing Disassembly

The front brake drums ride on two bearings. The outer bearing is held in place by the spindle nut and

Once the brake drums have been removed, you can inspect the condition of the wheel brake assemblies. This front wheel brake setup from a 1950 Ford F-1 is similar to what you will see on your truck. Actually this truck's brakes are in pretty good condition. Some surface rust appears on the backing plates and shoes, but the springs are intact, the assembly has enough lining material so that the shoes are not wearing against the drums, and no leakage shows from the wheel cylinders.

washer and comes out when you remove the drum. The inner bearing is located at the rear, or inside, of the drum's center opening. This can be removed using a special wheel bearing puller, which is a hook-shaped tool that looks like a miniature crowbar, or by tapping on the inner race with a long punch inserted through the hub.

A speedier method for removing the inner bearing, and one commonly used by mechanics, is to replace the nut and washer on the spindle, then slip the brake drum over the nut so that the hub rests on the spindle and pull the drum toward you in a sharp, downward jerk that forces the bearing cage against the spindle washer. This sharp tug will pop the bearing free nearly every time.

Occasionally it may be necessary to rotate the drum 180deg and jerk it against the spindle nut

Although certainly a candidate for rebuilding, this rear wheel brake setup, also from a Ford F-1, appears to have been in working condition.

You will probably want to sandblast and refinish the brake drums and backing plates. The best way to repaint the backing plates is to suspend them by wires from your shop ceiling.

Advice for Converting 1928-1938 Fords to Hydraulic Brakes
By Dave Moyer

I have converted a few 1935-38 Fords to hydraulics, and this is exactly what is needed to make the conversion: The parts required (starting at the front) are
- 1937-41 spindles #78-3105
- kingpin and bushing kit for 1937-41 spindles #78-3105
- 1939-41 backing plates with all parts (wheel cylinders, brake lining, etc.)
- 1939-48 front brake hoses
- 1937-39 front hub and drum assemblies
- 1939 brake and clutch pedal assembly, if possible (this is almost a bolt-in)
- cross-member cut slightly for master cylinder clearance
- 1939-48 master cylinder
- 1939-41 rear backing plates, complete
- 1936-39 rear hub and drum assemblies
- 1939-48 rear brake hose
- 1939 emergency brake cable (save one brake rod to use for the emergency brake front rod—cut and thread to length)
- left junction block
- master cylinder block, wishbone bock, and clips

Cut the brake tubing to the following lengths, and it will be correct when it is double flared:
- right front, 60 11/16in
- left front (master cylinder to left front connector block), 38 9/16in
- right rear, 70 3/8in
- left rear, 63 7/8in
- master cylinder to rear brake hose, 19 7/16in

These lengths are for 1939 Ford passenger cars or pickups and are also correct for 1935-38 Fords. Most dealers can supply the correct brake tube and fittings. Do not use copper tubing; it is unsafe and illegal.

A few notes: The 1936 front hub and drum assemblies will not fit the later spindles. The brake light switch hits the battery; put a 90-degree fitting on for clearance. The front emergency brake rod rattles on the cross-member and wishbone; slip a piece of hose over the complete length of the rod.

On attaching the brake tube to the wishbones, use plumber's tape from the hardware store and drill the wishbones, or use universal hose clamps around the tube and wishbone. Either way, wrap black plastic tape around the tube to insulate it from metal-to-metal contact. Other than this, the conversion is a bolt-on deal.

again. Although this method usually doesn't damage the seal, on a twenty- to fifty-year-old vehicle, these seals are usually dried out and should be replaced.

Brake Drum Inspection and Refinishing

It's not a good idea just to assume that the brake drums can simply be cleaned, refinished, and reused. Instead the drums should be carefully inspected for cracks, warpage, and inadequate wall thickness. If a drum has become cracked from excess heat build-up, the crack will usually show on the lining sweep area. Warpage can be checked by laying the drum face-down on a flat surface to see if it lies flush against the surface. On a 1953–56 F-100 half-ton pickup, the inside drum diameter should not exceed 11.06in.

Brake drums on these trucks have an original inside diameter of 11in. Maximum brake drum turning limits are found in the service manual. Measurements this fine should be taken with a brake drum gauge. A brake shop will have this measuring tool and will be able to tell you if the drums have enough metal to be turned, or if they will need to be replaced. You should also replace cracked or warped brake drums.

Brake drums can be prepared for painting by

Look for Set of 1939 Drums
By Mike Schmader

You will need to locate a full set of 1939 drums, preferably ones that haven't been turned; up to 60,000 miles is max, and drums with higher mileage are subject to brake fade under extreme driving. The next step is mounting the backing plates. If you get the whole brake set, including the master cylinder, so much the better. If not, 1939-48 setups were interchangeable, the difference being that the 1939-42 versions had adjustable anchors. Retailers used to sell a kit to fit between the backing plate and the spindle lip, and a Ford V-8 (any year) valve seat to fit between the inner bearing and the butt on the spindle. I am sure spacers are a must for the Model A through 1934, maybe 1935. As I recall, the rears didn't need spacers.

The problem of the master cylinder should be approached by using the stock Ford master. It worked very well even without the power assist because it was engineered for the overall weight, drum, and shoe area. Mounting a swing pedal on the firewall detracts from the original looks, and the firewall doesn't have enough "beef" to support the leverage of a swing pedal without some additional support. Much simpler, and original looking, is to mount the master cylinder behind the crossmember. This is done by removing the crossbar brake rod actuator or doing whatever is required or easiest to mount or fabricate a mount. Chances are, the existing linkage can be modified to actuate the master.

The real important phases are the location and mounting of the flex lines, and any brakes one might find today will require a brake overhaul.

If the drums are smooth, try to live with them. If not, find a "friendly" brake shop and get them turned the very minimum needed to clean them up. On 1939-42 backing plates with adjustable anchors, turning the drums can be compensated for by adjusting the anchors. This almost always requires a tool for transposing the diameter of the drum to the diameter or the shoes. The 1946-48 backing plates had non-adjustable anchors and required oversize shoes, plus the same tool to transpose the shoe-drum relationship as mounted. Arcing the shoe to the drum in a brake shop while the car is at home on jacks will result in a spongy or worse brake, leading one to think it wasn't bled properly.

When mounting the front flex lines to the frame, make sure they don't rub on the wheel on lock-to-lock turns. The rear one mounts next to the U-joint.

Changeover Works Well
By Herb Ward

I just completed a conversion to hydraulic brakes on my 1935 Ford pickup. I used 1939 rear backing plates with my 1935 rear wheel cylinders, shoes, and so forth. For the front, I used 1939-48 spindles and kingpins, and 1940 drums, backing plates, wheel cylinders, shoes, and so forth. It all fit on my 1935 Ford axle, but some changes in the tie rod and drag link may be required, as some of these parts are above and some are below the right spindle steering arm.

For the parking brake, I used a reproduction brake cable to the wheels, with about a 3ft-long Chevy wheel cable and housing tied to the original brake handle and linkage, and some homemade connectors. It works well. I left the brake rods in place and anchored the rears to the backing plates. It's hard to tell it's hydraulic without crawling underneath it.

I used a 1939 brake and clutch pedal assembly and hanger, and a 1939-48 master cylinder and stoplight switch. With some careful reworking of the frame, it will fit right in. The 1939 clutch and brake pedal assembly is the only one that will work. I like it better than swing pedals, as clutch operation is easy. The original assemblies cost $75 to $100 where I live, and usually need rebuilding. My changeover works great, but stopping is limited by the traction of the 6x16 tires.

sandblasting or wire brushing them and treating any remaining surface rust with an oxidation neutralizer like Fertan. Acid derusting should never be used on brake drums because dipping a brake drum in an acid bath can make the metal brittle, possibly causing the drum to crack.

Wheel Cylinder Disassembly

Wheel cylinders fail—that is, leak fluid or become gummed up and sluggish—owing to contamination of the hydraulic fluid. Unlike those in newer cars and trucks, the hydraulic brake systems of older trucks are not sealed to atmospheric moisture. Prior to the sealed hydraulic brake systems that were introduced in the late sixties, brake assemblies had a small vent hole that allowed air to enter as the fluid level dropped in the master cylinder reservoir. The vent hole existed to prevent a vacuum from developing as fluid transferred to the wheel cylinders to compensate for brake lining wear.

Along with air, moisture also entered these vented brake systems. The standard DOT (Department of Transportation) 3 brake fluid found in most cars and trucks sucks up moisture like a sponge. Once inside the brake system, this moisture rusts brake lines and wheel and master cylinder linings. In time, the brake fluid becomes gummed up with water and rust, causing the brakes to perform sluggishly, if at all. The wheel cylinders on an older truck typically show the damage of moisture inside the hydraulic brake system.

To determine whether the wheel cylinders on your truck are still in good condition or need to be replaced, you first have to remove them from the backing plates, then disassemble them.

Sometimes the bolts that hold the wheel cylinders to the backing plates are rusted frozen. Where this is the case, you can heat the cylinder casting with a torch—but first you should pull off the rubber boots that slip over the ends of the cylinder, and remove the internal parts. Otherwise you risk igniting the rubber cups and boots as well as the brake fluid. Simply peel back the rubber dust boots, then push the internal parts out one end of the cylinder. The parts that make up a working wheel cylinder include a spring, two rubber cups, and two pistons, which are small metal cylinders.

With the wheel cylinders disassembled and cleaned, you can point them toward a light source and inspect the condition of the bores. Chances are, you'll see the cylinder lining well scarred with pits. If the pits are shallow, a smooth lining surface can be restored by honing. In this process, the bore is enlarged a few thousandths of an inch with a grinding stone. Honing is something you wouldn't do at home but would have done by a machine shop.

With trucks of the age we're dealing with, pits in the brake cylinder linings are likely to be too deep to be removed with honing. My advice is to ignore honing and if any pitting is present, either purchase new wheel cylinders or have the old cylinders relined. Smooth bores are necessary to prevent fluid leakage at the wheel cylinders.

With the wheel cylinders removed from the backing plates and cleaned, you can check their condition by pointing them toward a light source and looking carefully at the bores. If the cylinder linings are scarred with pits, the wheel cylinders will need to be either rebored or replaced.

If the wheel cylinders didn't show any pitting, or if you decided to have them rebored, you will need to install a wheel cylinder rebuild kit. These kits typically consist of new springs, seals, and boots, but usually require that you reuse the old pistons.

Master Cylinder Disassembly

If pitting is found in the wheel cylinders, chances are, the master cylinder bore will also be corroded. On trucks with through-the-floor pedals, the master cylinder can be found under the floorboard behind the brake pedal on the driver's side. On trucks with firewall-mounted swing pedals, it can be found on the firewall inside the engine compartment.

Once the master cylinder is removed from the truck, its disassembly is a simple process of prying off the rubber boot and removing the snap ring that holds the piston, seals, and compression spring inside the casting. While prying off the snap ring, hold the stop plate to prevent the piston and spring from flying out of the housing when the ring pops free.

Wiping the master cylinder bore with a clean cloth and holding it up to the light will show you whether or not the bore is pitted. If pitting is visible, add the master cylinder to your parts-needed list.

Parts

The brake system parts list will consist of the following items:
- front and rear brake shoes
- brake shoe clips
- brake shoe retainer springs
- flex hoses
- wheel cylinder rebuild kits or new wheel cylinders
- brake lines
- master cylinder
- brake drums, depending on their condition

Many, if not all, of these parts may be available from a local auto parts store. NAPA auto parts stores maintain a large parts inventory for older vehicles, and if the items are listed on its computers, the local store can have the parts you need within 24 to 48 hours.

If the wheel cylinders do not show pitting, you should still install a rebuild kit; this is a less expensive option than buying new wheel cylinders. If the wheel cylinder bores are pitted, you need either to have the cylinders sleeved and install the rebuild kit or to buy new wheel cylinders. Resleeving the cylinders has little cost advantage versus buying new ones. The advantage of resleeving is that resleeving services install noncorrosive liners. If you decide to have the wheel cylinders resleeved, some extra time will be required to install the rebuild kit.

On trucks built before 1968, a rupture anywhere in the lines will cause the loss of all brake action. Therefore the only wise approach is to replace old brake tubing whether you are restoring or rebuilding the hydraulic brake system. (In a frame-up restoration, the brake tubing would be replaced as a matter of course.) Instructions for cutting new tubing, bending it to match the original, and forming correct double flares on the ends are provided later in this chapter, under the "Brake Line Replacement" section. Brake tubing and fittings are available at most auto parts stores.

Wheel Cylinder Rebuild

If you are installing new wheel cylinders, you can skip this procedure. You will rebuild the wheel cylinders if you either have had the cylinders relined or are working with used wheel cylinders that have absolutely clean, smooth bores.

Before installing the rebuild kits, refinish the castings. Some relining services will bead blast the castings and send them back in ready-to-paint condition. The castings can also be prepared for painting by wire brushing them and treating the metal with a rust neutralizer like Fertan. Aerosol spray paints work well for painting these small parts. When spraying the casting, mask off the ends of the cylinder to keep paint out of the cylinder bores.

Wheel cylinder rebuild kits typically consist of new springs, seals, (also called cups), and boots, but require that you reuse the old pistons. It's important to be sure the pistons are smooth and free from

Before the wheel cylinder rebuild kit is installed, the bore should be wiped clean with a lint-free cloth, then coated with a special lubricant (sometimes included with the kit). Petroleum-based lubricants must never be used on any of the hydraulic brake system's internal parts.

To rebuild the wheel cylinder, the spring is inserted into the cylinder, and then the seals are moistened with a special lubricant and fitted against each end of the spring. A vise is useful in this operation, to hold the wheel cylinder firmly in place.

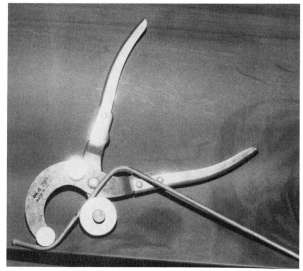

New brake line tubing is bent to match the shape of the old brake lines using a bending tool. To make the angles, the tubing is placed between the guides and the forming die, and the handles of the bending tool are squeezed together.

scratches. To prepare the cylinders for reuse, first clean them in solvent, then scrub the outer cylinder area with superfine 0000-grade steel wool. Follow this by sanding with 240-grit sandpaper to smooth any scratches, and then polish the surface with 600-grit automotive sandpaper. After sanding, wash the pistons in soapy water to remove all sanding residue. If the pistons are badly scratched or pitted, they will need to be replaced.

Before reassembling the wheel cylinder, wipe the bore clean with a lint-free cloth, then coat the cylinder with a special brake system lubricant. Petroleum-based lubricants must *never* be used on any of the hydraulic brake system's internal parts. Wheel cylinder rebuild kits will typically contain small vials of the special lubricant to be used for this purpose. If yours does not, the lubricant can be purchased at an auto parts store.

Now the spring is inserted into the cylinder. Next the seals are moistened with lubricant and fitted one against each end of the spring. The inner lip of the dust boot is designed to fit into a groove in the outer end of the piston. You will find it easier to fit these two parts together now, rather than after the pistons have been pressed into the cylinders. Putting a few drops of the special brake system lubricant on the dust boot lip will help the rubber slide into the groove in the piston.

Also lubricate the pistons before sliding them into the cylinder. Now place a few drops of the special lubricant on the outer lips of the dust boots and snap them over the ends of the casting. Thread new bleeder screws into their openings on the back side of the casting (bleeder screws are available from most auto supply stores), and seal the brake line openings with tape to prevent dust from entering the cylinders—and this job is finished.

Brake Line Replacement

To have a pattern for the new brake lines, try to keep the old tubing as intact as possible as you remove it from the truck. If you spray the connections with penetrating oil and use a wrench with a semibox end—to get a good grip on the fittings—you should be able to unscrew most, if not all, of the connections and have good patterns for the new lines. The long lines to the rear brakes will probably be fastened to the frame with clips. These will either slide off the frame or need to be unbolted.

With the old lines removed, the next step is to measure each line with a metal carpenter's tape. Be sure the tape follows each bend and curve. Now mark down the length of each line, and take these measurements to an auto parts store. There you will find new brake line tubing in a range of precut lengths. If the precut lengths don't match those on your list, purchase slightly longer lengths, then cut them to the measurements for your truck and reflare the ends. Even if the precut lengths match your measurements, it may still be necessary to cut and reflare the ends on the tubing to be used for the front brakes, in order to fit protective steel coils over those lines. At the factory, steel coils were placed over the portions of the front brake lines that pass through the fender shields, to prevent possible metal-to-metal contact from cutting the line.

The new brake tubing is bent to match the shape of the old brake lines using a bending tool, available from most auto parts stores or a specialty tool suppler like The Eastwood Company. This tool looks like a large pliers fitted with two round guides and a forming die. To bend brake tubing, a die matching the tubing diameter—1/4in, in this case—is installed on the bender. The tubing is placed between the guides and the forming die, and then the handles of the bending tool are squeezed together. The tighter the handles are squeezed, the sharper the bend.

Tubing can be bent at any angle. The only trick to using the bending tool is to position the tubing so that the spot where you want to make the bend is directly over the forming die. To get used to how the bending tool works, make several practice bends with a spare length of tube.

After all the bends have been made in one of the brake lines, place the new line beside the old one to make sure you have created an exact match, then mark where the new line is to be cut. A tubing cutter, not a hacksaw, is used to cut brake tubing.

On the short lines to the front wheels, protective wire coils are now slipped over the end that has been cut—if your truck used these coils originally.

Now you are ready to flare the freshly cut ends. It is very important that the flaring operation be done correctly. If it isn't, brake fluid will leak from the connections, causing brake failure. To make the correct double-wall flares requires a flaring tool, holding bar, and crimping die for 1/4in-diameter tubing. These items can be purchased from an auto supply store and are sometimes available from a tool rent-all.

The process for flaring brake tubing consists of these four steps:

1. Clamp the tube tightly in the holding bar so that a length of tube equal to the larger-diameter lip of the crimping die sticks out above the bar. Dip the end to be flared in brake fluid. (Lubricating the end helps make a proper flare.)

2. Place the crimping die over the end of the tube, then fit the flaring yoke over the bar. Now screw the yoke down until the die seats against the bar.

3. Loosen the yoke and remove the crimping die.

4. Tighten the flaring yoke again until the tube is fully flared.

After completing these steps, remove the tubing from the holding bar and carefully examine the flare

Flaring the ends of the new brake lines is a critical operation in the brake overhaul procedure. If the flaring operation is not done correctly, brake fluid will leak from the connections, causing brake failure. Making the correct double-wall flares requires a flaring tool, holding bar, and crimping die for 1/4in-diameter tubing.

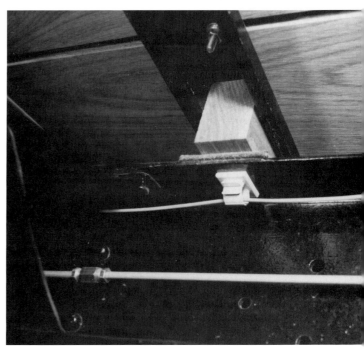

After the lines are bent and flared, they will be installed on the chassis. Since brake tubing is bare steel, the new lines will rust unless they are treated. To preserve that factory-new look, many restorers paint the brake lines with clear lacquer. Bruce Horkey

for cracks. If you spot a crack in the flare, you will have to cut the tube again and make a new flare. In most cases, this will not require bending a new line, since you can usually stretch the line to the needed length by slightly reworking some of the bends and curves. If several reflaring attempts fail, however, it will probably be necessary to cut and bend a new line.

After all the lines are bent and flared, they will be installed on the chassis. Since brake tubing is bare steel, the new lines will rust unless they are treated. To preserve that factory-new bare-steel look—brake lines were not painted when the truck was assembled—clean the tubing with metal prep; scour the bare steel with superfine 0000-grade steel wool, after first wrapping masking tape over the ends to prevent steel wool splinters from entering the tubing; and then paint the tubing with clear lacquer.

When installing new brake lines, it is important that all connections be tight. If you are not hooking the lines up to the wheel cylinders at this point, leave the ends of the lines taped to prevent dust and other contamination from entering.

Brake Reassembly

On a frame-up restoration, the front suspension will have been checked and rebuilt, and all components cleaned and painted. The rear axle assembly will also have undergone the same thorough examination and been rebuilt as necessary. With the rebuilding approach, the brakes will most likely have been overhauled separately. In either case, the assumption here is that the backing plates

As an alternative to the procedure described in the text, the brakes can be reassembled on the backing plates, then installed on the truck.

have been bolted back onto the axle flanges.

The steps in the brake reassembly sequence are to mount the wheel cylinders on the backing plates, connect the brake tubing to the wheel cylinders, and replace the brake shoes. Before mounting the brake shoes, it is important to wipe a coating of light lubricant on the anchor pins and to oil the adjusting mechanism. Skipping this step can result in squeaky brakes that become hard to adjust. On the rear brakes, you should also lubricate the parking brake cables.

Next connect a set of shoes to the adjuster by installing the spring at the bottom of the shoes, on 1948 and later trucks. On pre-1948 trucks with hydraulic brakes, the shoes connect at the bottom with an anchor plate; only one return spring is used, and it attaches near the top of the shoes. If one of the linings is shorter, this is the primary shoe and goes toward the front. Make sure the star wheel on the adjuster is over the adjusting hole in the backing plate. Now attach the shoes to the backing plates with the hold-down pins and springs. Check to make sure the shoes have seated in the slots in the wheel cylinder connecting links.

On rear brakes, connect the parking brake cable. Place the anchor pin plate over the anchor pin; this plate secures the top of the shoes. Next clip the brake return springs through the holes on the shoes, then stretch the spring from the primary shoe over the anchor pin. Now pull the spring from the secondary shoe over this same pin. You will find this job much easier to do if you use a special brake tool than if you attempt to spread the springs with pliers or a screwdriver.

Before replacing the brake drums, pull the shoes away from the backing plates and tap them a couple of times to make sure they are seated on the wheel cylinder and adjuster. Also make sure the adjusters are turned all the way in.

At this point, you can slide on the rear drums. Remember to replace the three 5/16in bolts that hold the rear drums to the axle flange on F-1 and later trucks.

Wheel Bearing Packing

Before replacing the front brake drums, you will need to repack and replace the wheel bearings. Packing wheel bearings can be a messy job. When you're finished, your hands are completely gooped with grease and you probably have some grease on your clothes as well.

The packing method I learned from my father is to scoop up a gob of wheel bearing grease and place it in the palm of one hand, then stroke the bearing—which has been cleaned in solvent and allowed to dry on its own (do not dry bearings by spinning them with an air gun)—through the grease with the other hand. After two or three passes, Dad would tap the bearing against the heel of his hand to pack the grease; then he would repeat the scooping and tapping steps until grease oozed out around the bearing rollers. This process isn't complicated, but it is messy. An easier method may also be used.

Aerosol wheel bearing packers that pump grease into the bearing—thereby eliminating the messy hand packing steps—are available at most auto supply stores and discount marts. To use the aerosol packer, just place the bearing in a funnel-shaped clamp, supplied with the packer; insert the aerosol nozzle into the packer; and press down on the grease container. In less time than it takes to say, "Peter Piper packed a pair of bearings," grease will be oozing out of the clamp and the bearing will be packed and ready for installation.

The inner wheel bearings are installed first and are held in place by a seal ring that presses into the brake drum. When these seals are not available from a local auto supply store, replacement seals can be mail-ordered from a vintage Ford truck parts supplier. The seals are seated by being tapped into the hub with a plastic hammer. Now the front brake drums can be slipped onto the spindle and over the linings. If the drums won't fit, pull them back off, tap the shoes with the heel of your hand to seat them tightly against the wheel cylinder, turn the adjusters to make sure the shoes are fully retracted, and try again. With new linings, the fit may be snug, but the drums should slide over the shoes, with perhaps some turning and tapping.

Now the outer wheel bearings can be packed. When this is done, the bearings are slipped onto the spindle and pushed into the bearing cone in the brake drum hub. A washer fits between the bearing and the spindle nut. Tighten the spindle nut until the bearing drags as the drum is turned, then back the nut off until the drum spins freely. Lock the spindle nut in this position with a cotter key, and spread the key to hold the nut in place.

Brake Adjustments

Adjustments to brakes on Ford F-1 and later light trucks are made by reaching through the access hole in the bottom of the backing plates and turning the star wheel. Although a screwdriver can be used for this operation, a special brake-adjusting tool, shaped like a lazy Z, works best. The brakes should be adjusted by moving the shoes toward the drums until a slight drag is felt, then backing off the adjuster wheel ten or twelve clicks until the drums turn easily. Adjustments at all four wheels need to be as uniform as possible to prevent one or more of the wheels from locking under panic braking and to keep the shoes from dragging.

The earlier-style brakes adjust by turning the cams that are located at the three-o'clock and nine-o'clock positions on the rear of the backing plates. Adjusting this style brakes is somewhat more

Two methods can be used to repack wheel bearings. The old-fashioned method is to scoop up a gob of wheel bearing grease, place it in the palm of one hand, then stroke the bearing through the grease with the other hand.

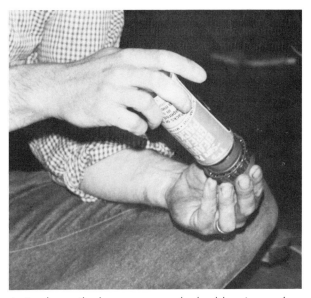

A simpler method uses an aerosol wheel bearing packer. In a matter of seconds, the bearing is packed and ready for installation.

complicated than adjusting the F-1 and later type. The drums on these trucks have a feeler gauge slot in the front. To adjust the brakes, rotate the drums until the feeler gauge slot is in front of the bottom of the secondary, or rear, brake shoe. Now insert a 0.007in feeler gauge through the slot and move the feeler up the secondary shoe until it wedges between the shoe and the drum. Then loosen the anchor pin nut for the secondary shoe and turn the secondary shoe anchor pin until the feeler gauge can be inserted between the shoe and the drum at a point 1-1/2in up from the bottom of the shoe. The anchor pin nuts are located at the bottom of the backing plate; the anchor pin for the secondary shoe is the rear of the two nuts.

Next rotate the drum until the feeler gauge slot is in front of the upper end of the secondary shoe. Now insert a 0.01in feeler gauge through the slot and move the gauge down until it wedges between the shoe and the drum. Next turn the adjusting cam until the shoe drags against the drum. Then turn the anchor pin until the clearance between the brake shoe and the drum at a point 1 1/2in down from the top of the shoe is 0.01in. This done, remove the feeler gauge and torque the anchor pin nut to 80lb-ft to 100lb-ft.

Repeat this procedure on the primary, or front, shoe and at the other wheels.

Brake Line Filling

Although manufacturers fill hydraulic braking systems with DOT 3 polyglycol brake fluid, many old-car and -truck owners are switching to DOT 5 silicone brake fluid. Unlike the polyglycol fluid, silicone brake fluid does not attract moisture. Another benefit is that silicone fluid lubricates and helps preserve rubber brake system parts such as wheel cylinder seals and flex hoses.

The disadvantage of silicone fluid is that it has a tendency to destroy hydraulically actuated stoplight switches—the type used on Ford light-duty trucks with master cylinders mounted under the floor. Since these switches are inexpensive items and installed quite easily—just pull off the brake light wires, unscrew the old switch from its mount on the master cylinder, thread the new switch back in, and slip the wires back on their connectors (the process takes maybe 15 minutes)—it seems preferable to replace the stoplight switch, when necessary, rather than go through the hassle of flushing and purging the brake system on an annual basis to remove contaminated brake fluid—as should be done when polyglycol is used.

To refill the brake system, simply pour the brake fluid into the master cylinder through a funnel. Stop when the reservoir is full.

Caution: If you decide to use DOT 3 polyglycol fluid, fill the brake system from a new, unopened can. A container of brake fluid that has been sitting on your shop shelf with the cap seal broken will have already absorbed enough moisture to contaminate your truck's freshly rebuilt brake system.

Hydraulic Brake System Bleeding

Whether you use DOT 3 or DOT 5 brake fluid, before the truck can be driven, trapped air needs to be bled, or purged, from the brake lines. This can be done in several ways. The simplest requires two people and takes little more than one half hour—assuming all goes well. The helper will pump the brake pedal to build up pressure, and you will bleed the air from the lines by loosening the bleeder screws at each of the wheels.

Begin by making sure the master cylinder reservoir is topped up with brake fluid. When the

When you are bleeding the brakes, a length of plastic or rubber tubing should be slipped over the tip of the bleeder screw to prevent brake fluid from squirting on the chassis and running down the backing plate. If you have decided to refill the hydraulic brake system with DOT 3 fluid, this procedure is especially important, as any DOT 3 that contacts refinished brake or chassis parts will eat off the paint.

fluid reservoir is full, screw the cap on tight and have the helper pump up the brakes—tell him or her to push the brake pedal down several times in quick succession until braking action is felt. While the assistant holds a foot on the pedal, you will proceed to the wheel farthest from the master cylinder—the right rear wheel—and loosen the bleeder screw to the wheel cylinder; the bleeder screw threads into the back side of the wheel cylinder and is reached from underneath the truck. To keep brake fluid from squirting on the chassis and running down the backing plate—polyglycol fluid will eat off paint, silicone fluid is harmless—fit one end of a length of plastic or rubber tubing over the tip of the bleeder screw. Place the other end of the tube in a can or jar to catch escaping fluid.

As the bleeder screw is turned open, air will escape from the brake lines. When this happens, your assistant will feel the brake pedal sink slowly toward the floor.

As soon as the fluid runs clear and no longer contains bubbles, turn the bleeder screw shut. Now ask your assistant to pump up the pedal again and continue to hold pressure on the pedal. With your assistant continuing to press on the brake pedal, open the same bleeder screw again to make sure fluid from that line runs clear and without bubbles. Be sure to warn your assistant not to release the brake pedal once it sinks to the floor, until you say to do so. Releasing the brake pedal before the bleeder screw is closed will allow air to be sucked into the lines, requiring that the bleeding process be done all over again.

Once air is purged for the line to one wheel, refill the master cylinder, then move to the next wheel and repeat the bleeding process. Less than a quart of fluid should be required to fill and bleed a rebuilt brake system on a light truck. After all the lines have been bled, check all connections to make sure none leak.

Brake System Maintenance

To keep your truck's hydraulic brake system functioning properly, check the fluid level of the master cylinder periodically and watch for leaks at the brake line connections and from the wheel cylinders. Serious wheel cylinder leaks will show up as streaks on the inside of the tires. If polyglycol fluid has been used, the brake system should be flushed and refilled on an annual basis. On more modern trucks with self-adjusting brakes, the only other maintenance step would be pulling the brake drums and checking for lining wear at 30,000mi intervals, or thereabouts. On trucks without self-adjusting brakes, add brake adjustment to the fall or spring truck care session.

A properly rebuilt and maintained hydraulic braking system will give your truck more-than-adequate stopping power and ensure driving safety for you and your passengers.

Chapter 10

Metal Repair

Unless your truck has spent its life in the dry climate of the western plains or the desert Southwest, or has been exceptionally pampered and well-cared-for, a major phase in the process of restoring or rebuilding it will be repairing rusted and dented metal. The first step in this process is stripping off old paint and surface rust, as described in chapter 5. When the restoration approach is followed, the truck is typically disassembled during the derusting and stripping process. With rebuilding, the fenders, doors, and box may be removed, but the cab is generally left on the frame.

You can test for plastic filler that covers earlier rust and dent damage, by running a magnet over suspect areas like the fenders, lower door sections, and rear cab corners. But sandblasting or having the metal chemically stripped tells the full story.

Sometimes rust damage will look rather minimal: a small hole surrounded by tiny pinholes. Truth is, the rust area you'll need to replace is much larger than it looks. The easiest way to find out the extent of the rust damage is to jab a small-bladed screwdriver into suspect areas. Don't be upset if the screwdriver pokes through the metal. You haven't caused any problem that didn't already exist. If the metal had been solid, the screwdriver wouldn't have ripped through. All you've done is shown where the metalwork needs to be done. On pickup trucks, potential rust areas include the bottom of doors, the cab corners, and the cowl panels just past the rear of the front fenders, as well as the bottom of the pickup box supports.

If the rust damage to your truck is extensive, you'll be time and money ahead to look for better parts. For F-1–series and later Ford trucks, rust-free cabs and doors are still relatively plentiful in states south and west of the Rust Belt. Patch panels for repairing common rust areas such as the bottom of doors and cab corners are available for most years of Ford trucks. Installing patch panels is a rather easy job that an adventuresome hobbyist can do, provided she or he has a welder and a few other specialty tools. You will find a presentation of how to install patch panels later in this chapter.

Dents are straightened by pounding against low and high spots with a bumping hammer while pressing against the other side of the metal with a body dolly. The dolly spreads the hammer's force, allowing the metal to move back into shape without leaving a surface pocked with hammer marks. The Eastwood Company

Dent Repair

Since trucks are working vehicles, dent damage is common. Although dents can occur anywhere, the most common places are in the fenders and the cab rear panel and roof. Fenders can be straightened most easily if they are taken off the truck. Straightening dents in the cab back panel and roof requires removing the seats and the headliner.

It's best to work out dents with the metal at air temperature. In some situations, however, severe dent damage in the heavy-gauge sheet metal of older trucks will require heating the metal with a torch. Although heating makes the metal easier to straighten, it also increases the chances of stretching the metal. When metal becomes stretched, you can't straighten the dent because no matter how much you work the metal, you still have a bulge that you can't figure out how to flatten. To smooth the dent, this bulge will have to be shrunk.

Dents are straightened by pounding alternately against low and high spots with a special "bumping" hammer while pressing a body dolly against the other side of the metal. The dolly spreads the hammer's force, allowing the metal to move back into shape without leaving a surface pocked with hammer marks.

Much of the skill of "metal bumping," as dent straightening is called, comes from developing the feel of how hard to strike the metal with the bumping hammer to work out the dent. The best way to learn this feel is by spending some time practicing on pieces of scrap metal. This could be done by working on a beat-up fender from a scrap yard or on dents you have pounded into a piece of metal that is the same gauge as the body panels on your truck. When you've acquired the knack of working out dents with the bumping hammer and dolly, you can begin to straighten the dents on your truck.

Bumping hammers and dollies come in a variety of shapes and sizes. The hammer heads typically have two striking surfaces, which may be rounded, squared, or somewhat pointed. Each shape is designed for straightening metal in slightly different situations: working with contoured or flat panels, or removing dents from corners or moldings, for example. You don't need to own the whole gamut of body hammers to work the dents out of your truck's fenders, but a couple of different hammers will give you some flexibility in picking the tool that feels right for the job. Likewise body dollies come in a variety of shapes to match various metal contours, and here again, having a couple of dollies to pick from will give you a chance to find one that matches the shape—curved or flat—of the metal you are straightening.

The first step in straightening a dent is to move the metal back to its original contour with a bumping hammer and dolly. Rather than try to smooth the dent with a few sharp blows, work it out gradually. Manipulate the hammer and dolly from the outside of the dent toward the center. Your goal isn't just to straighten the metal; you're also trying to

Minor dents and surface imperfections can be filled with a skim coat of body filler.

keep it from stretching. If you stretch the metal in the process of working out the dent, you will wind up with a bulge that you won't be able to flatten no matter how much you pound at it with your hammer and dolly. To get rid of a stretch bulge, either you have to shrink the metal with a special shrinking hammer and dolly that have serrations on their facing surfaces, or you have to heat the bulge with a torch, then quickly cool the metal with water. This heating and quenching method works well on older, heavy-gauge metal body panels but should not be used on modern high-strength steel panels. (For instructions on the metal-shrinking process, refer to the "Metal Shrinking" section later in this chapter.)

If you're able to smooth the dented area so that only minor surface imperfections remain, the hammer marks can be smoothed with a body file and primer can be used to fill any remaining blemishes. More likely, however, a skim coat of body filler will be needed to cover hammer marks and spots that you weren't able to bring completely back to the original shape.

Rust Repair

Replacing rusted metal requires a welding outfit and welding skills, plus a few specialized tools. Although tin snips or a hacksaw can be used to cut away rusted metal, a nibbler or cutoff tool makes the job much easier. A patch will need to be cut to fit the rusted area, and a panel nibbler also makes this

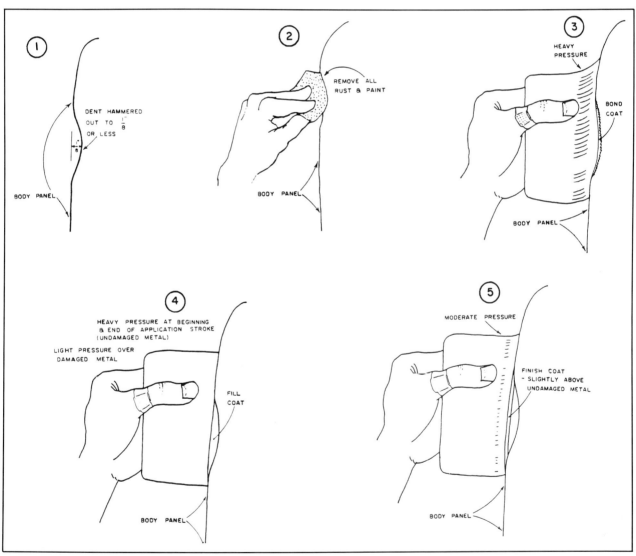

To do a repair job with plastic filler, make sure the metal is bare and clean, then scoop some from the can and mix it with a hardener. The mixture will take on a new, consistent color as the hardener is mixed in. Spread a thin coating of filler over the repair area with a plastic applicator (available where you buy the filler). After the filler hardens—which takes about 20 minutes—smooth the coating with a grater or a body solder file. Mike Barnes

job easier. Other tools that are useful in the metal repair process include:
- panel flangers—used to form an offset flange so that replacement panels can be fitted flush with the surrounding metal
- crimping pliers—used to install door skins
- carbide burr—used to grind weld beads
- panel holding clamps—used to hold panels together for welding
- heat sink putty—used to prevent heat warpage

Patch panels can be installed using either gas or arc welding, but as the next section explains, wire welding—a form of arc welding—is the superior method for attaching patch panels because it creates the least risk of heat damage to the surrounding metal. If you are comfortable with arc or gas welding, you will be able to wire weld after a short practice time. If you've not had welding training or experience, you may want to consider taking a welding course at a skill center or technical college, in preparation for doing the metal repair on your truck.

Wire Welding

Both gas welding and the standard type of arc welding that uses flux-coated rods for the electrode create problems when used to repair sheet metal. Gas welding produces very high heat that is likely to warp the metal. Arc welding is difficult to do on thin-gauge sheet metal. Wire welding, which is a special form of arc welding, avoids both of these problems. It produces much less heat than gas welding and is an easy welding method to master. The disadvantage of wire welding was that the equipment was quite expensive and therefore beyond the grasp of many hobbyists. But wire welding outfits designed especially for hobbyist restorers are now on the market, and these sell for not much more than a good-quality gas welding setup.

The basic difference between wire welding and standard arc welding is that with wire welding, the filler material is very thin (0.03in) and is fed automatically into the weld area. The thin filler material—actually a strand of wire, hence the name *wire welding*—is just right for filling the thin seams in sheet metal repair, and since the wire is fed into the weld automatically, the operator needs only to

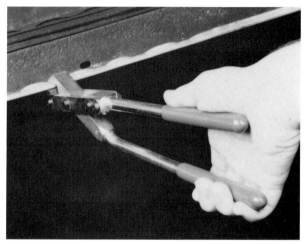

Another useful tool is a crimping pliers, used to fit a repair panel over a rusted outer door skin. The Eastwood Company

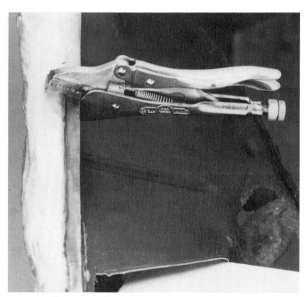

Among the specialty tools used with body repair is this panel flanger, which crimps an offset in the replacement panel so that it can be fitted flush with the surrounding metal. The Eastwood Company

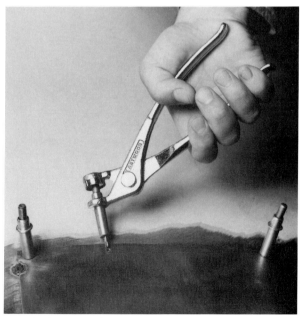

Panel clamps hold repair panels tightly in place for welding. The Eastwood Company

An inexpensive tool for arc welding patch panels is this stitch welder. The Eastwood Company

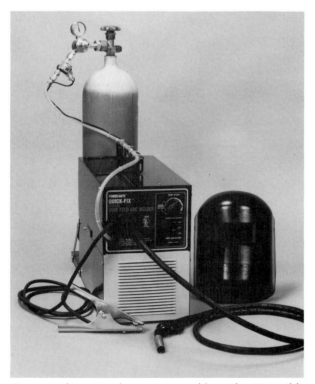

Because of its ease of use, a wire welder makes it possible to lay down smooth, even welds with a minimum of practice. Another advantage of wire welding is that it produces less heat than does gas welding and therefore reduces the risk of metal distortion. The Eastwood Company

concentrate on moving the welding gun along the work surface. With standard arc, or stick, welding, the operator has to keep the correct gap between the welding rod, or electrode, and the work surface while moving the rod to create the weld. Holding the rod the right distance from the metal while keeping it moving is somewhat of a trick, since the electrode is continually being consumed—therefore becoming shorter—as it forms the weld.

Welding Equipment

The hobbyist has two types of wire welding equipment to select from. The less expensive type uses special wire with a flux core. This wire is a little thicker, owing to the flux core, so the weld has a chance of burning holes in thin body metal. The more expensive type of wire welder hooks up to a cylinder of inert gas and uses this gas—rather than a flux core—to shield the weld. This type of welder is called a metal inert gas (MIG) setup. Because inert gas shields the weld, a MIG welder can use wire as thin as 0.024in. The price difference between the two types is not all that significant—about $100 plus the cost of the inert gas cylinder—so if you're buying a wire welder, you're advised to select the MIG type.

Because of its ease of use, a wire welder makes it possible to lay down smooth, even welds with a minimum of practice. Wire welding has the added advantage that it is equally easy to do in either horizontal or vertical positions. This means it is no more difficult to wire weld a patch to the rear cab corners than it would be to mend rusted metal on a door you took off the truck and placed on sawhorses to have a nice, horizontal surface to work on.

Welding Technique

The technique of wire welding consists of fastening the ground clamp to the metal being welded someplace near the work area, turning on the welder, setting the wire-feed control and heat dial, and then striking an arc with the electrode on the work surface and moving the wire-feed gun across the metal you are welding. About the only "trick" is getting used to the speed at which the wire feeds out of the gun. This can be mastered with a little practice.

A good way to become comfortable with wire welding is to practice welding small patches onto a piece of scrap sheet metal. An old hood or door panel works well. If you are using discarded car or truck body parts, you'll need to scuff off the paint with a dual-action sander in the area where you'll be doing the practice welding. When attaching patches, first tack weld the corners, then run the welding bead along the sides of the patch. This is basically the procedure you will follow when gas welding patch panels to your truck—except that you will take great care to avoid heat build-up that can warp the metal.

To keep heat from concentrating in an area, in attaching patch panels to your truck, you will weld a couple of inches along one seam, then release the feed button on the wire welding gun and reposition the gun at the opposite end of the seam and lay down another two or so inches of weld. Then you will go back and continue the first weld for another couple of inches, stop, skip back to the other end of the seam, and so forth. This technique of working along a seam in a back-and-forth fashion helps distribute the heat. On very large, flat panels, such as doors, heat build-up can also be controlled by using a heat dam made of asbestos putty sold through specialty suppliers like The Eastwood Company.

The problem most beginning wire welders have difficulty with is wire feeding out of the gun before they've moved the gun to the start of the seam and struck an arc. The extra wire will melt as soon as the arc is made, but it forms an amateurish-looking glob of metal at the start of the weld. A solution is to lift your helmet, with the welder turned off, and touch the electrode to the point where you want to begin welding. Then, as you lower the helmet over your face, you can turn on the welder to strike the arc. As soon as the arc appears, you can press the feed button and begin to move the gun along the seam. A steady cracking and hissing sound is a sign you are holding the gun the correct distance from the work surface and moving the gun at the proper rate.

Once the patch is completely welded, the seam needs to be ground smooth. Since wire welds are hard, this step is best done with a carbide grinder. If you're quite a good welder, no other finishing step may be needed before priming and painting. Usually, however, some filler—lead or plastic—will be needed to smooth the seam.

Wire welders for light-duty auto and truck bodywork are available from specialty tool suppliers like The Eastwood Company. If you have any arc welding experience, a little practice will make you a reasonably skilled wire welder. Instructions with the welder will guide you in setting the wire-feed and heat controls. If you have not had any welding experience, the best way to learn is to enroll in a welding class in an adult education program at a skill center or technical college. These classes typically cover gas, stick arc, and wire feed welding and offer you practice time in all three methods. Of utmost importance, in a class setting, you will learn to follow welding safety rules—the most important of which is always to wear proper eye protection.

Never weld without wearing a welding helmet. When looked at with the naked eye, the bright flashes from the weld can burn the retina. This damage is not repairable and can lead to blindness. If anyone else is in your shop when you are welding, caution him or her not to look at the weld. This is particularly important when children are around. Their natural curiosity will draw their eyes to the bright arc flashes. Always make sure children are out of the area before beginning to weld.

Cab Repair

Rust-prone areas on pickup trucks include the rear cab corners; the bottom of the doors; the front cowl section, at the rear of the front fenders; the pickup box supports; and the box sides on the 1957 and later Styleside trucks. This section explains how to repair rust damage to these areas using patch panels. You'll find that repair panels for the common rustout areas are available from specialty suppliers like Dennis Carpenter for most years of Ford trucks.

Where repair panels aren't available, they will need to be fabricated. Since most of the areas that need repairing on a pickup have relatively flat surfaces, it's possible to construct your own patches without having to form elaborate compound curves. Tools to help make your own patch panels are available from The Eastwood Company.

Damage Assessment

The first step in rust repair is to determine the extent of the damage. Usually a badly rusted section will look better than it is. The only sign of rust may

A good way to become comfortable with wire welding is to practice welding small patches onto a piece of scrap sheet metal.

be blisters in the paint. As mentioned earlier, jabbing a thin-bladed screwdriver into a paint blister is a quick way of finding out the condition of the metal. Don't worry about poking holes in good metal. If the metal is sound, the screwdriver won't punch through. If the screwdriver stabs a hole, dig around the area to determine the extent of the rust. Even when rust is evidenced by a gaping hole, you'll want to poke at the metal around the hole to find out how large to make the patch.

Hole Preparation

Once you've learned the extent of the rust, you can draw a pencil line around the rust area to show where to cut for the patch. To make the patch fit as neatly as possible, make the line as straight as you can. The cutout doesn't need to match the size and shape of the patch. In most cases, you'll cut the patch to fit the hole.

If you are cutting away extensive metal in the area of the floor, cowl, and door supports, you'll need to brace the cab. Otherwise, after you have finished the repairs, you may discover that the doors don't fit. For localized repair, like fitting patch panels into the rear cab corners, no bracing is needed.

An air-powered cutoff tool makes easy work of cutting away the rusted metal. Just guide the tool along the pencil line, and it will slice away the damaged metal in seconds. Be sure to wear safety glasses or goggles to protect your eyes against flying sparks and metal chips. If you don't own a cutoff tool, a hacksaw or tin snips can be used, but these leave jagged edges, which will have to be smoothed before the patch is installed.

In some areas, such as the cowl section, you may also need to cut tack welds that were made at

Among the most rust-prone areas on pickup trucks are the rear cab corners. Repair panels for these common rustout areas are available for most years of Ford trucks.

An air-powered cutoff tool makes easy work of removing rusted metal. If you don't own this tool, a hacksaw or tin snips can be used.

When new metal is needed in flat areas, such as the cab floor, a patch panel can be cut from sheet steel. Richard Matott

the factory to hold the sections of the cab together. The easy way to remove tack welds is with a spot weld cutter.

Patch Panel Preparation

Now that you have opened up the area for the patch, you need to cut the patch to fit the hole. An easy way to do this is to hold a piece of cardboard behind the hole and draw the outline of the hole on it. The cardboard can then be cut to the outline, and the shape transferred to the repair panel, or used to make a patch panel if a repair panel is not available or if you decide to save a few dollars and make your own.

Patch Panel Welding

Two methods can be used for welding in the patch. The easier method, and the one recommended for those without a lot of welding experience, is called a lap joint. To make a lap joint, you need to form a lip along the edges of the patch. When the patch is installed, this lip fits behind the outline of the hole. The lip allows a little leeway in cutting the patch and makes sure you have metal to weld to when attaching the patch.

The other, somewhat more difficult method is called a butt joint. Here the patch is cut so that it fits perfectly into the hole. When the seam is welded, the only finishing work is to grind the weld smooth. A butt weld shows superior artisanship but is more difficult to do, since the fit of the patch has to be nearly perfect.

The first step in making a lap joint is to cut the panel about 1/2in larger, all the way around, than the hole. Next you use a flanging pliers or a flange-rolling tool to form a lip on the edges of the patch. Now you are ready to fit the patch in place. For welding, the patch has to be clamped tightly to the body metal. This can be done with pop rivets (these can later be drilled out, and the holes filled) or with panel holders. Various welding methods can be used to fuse the patch to the body metal, but wire welding is preferred because of the advantages already mentioned. It is also advisable to weld or caulk the inside of the lap joint on the inside of the cab. If this joint is left open, moisture may penetrate the seam and blister the paint or rerust the metal. After the inside seam has been filled or caulked, the outside seam is smoothed by grinding.

To make a butt joint, the patch is cut so that it

With the rusted metal removed, you're ready to cut the patch and fit it in place.

Patch Panel Installation Steps

Here are the steps to follow as you do patch panel welding, and some tips about what can happen to metal as you weld.

1. Hold the patch over the hole and scribe around it.

2. Cut out the rusty section using a cutoff wheel, metal nibbler, or tin snips.

3. Cut the patch to fit the repair area. If you cut the patch 1/4in to 1/2in larger than the hole, weld warpage will be less and the edge can be crimped for a lap joint. ViseGrip hand crimpers can be used to form the overlap lip.

4. Grind all burrs, paint, and surface rust from the weld area.

5. Use a coating to prevent rust from forming under the lap joint and on the back of the patch. A recommended product is The Eastwood Company's Cold Galvanizing Compound.

6. Tack weld the patch about every 1in, then go back and fill between the welds 1/2in at a time. Alternate from one area of the patch to another to avoid warpage. Wire welding is the preferred method, but gas welding can be used if a heat dam is made with heat sink putty.

7. Apply filler and smooth the weld. Regardless of the welding method, very little filler should be required. The results are a permanent repair—much better than filling over rust, only to watch the filler bubble or fall off soon afterwards.

Lap joints or angles can be formed with handmade dies.

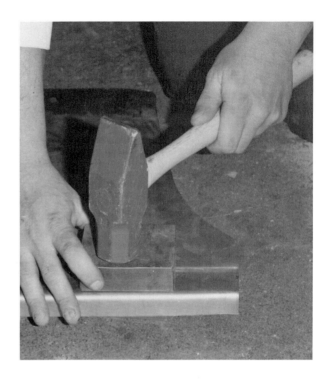

Two methods can be used for welding in the patch. The easier one is the lap joint, where the panel fits behind or under the edge of the surrounding metal.

fits precisely in place, then tack welded in several locations. If the welder has enough skill and the joint is a tight fit, the patch can be made so cleanly that the only finishing work is to grind the weld. When a gas welder is used to make the butt joint, a technique called hammer welding, which practically eliminates all finish work, can be employed.

To hammer weld a seam, you need to be able to reach both sides of the metal. Since gas welding will be used, it is important to be very careful about heat warpage. If you are working on a large panel like a door skin, a heat dam made from heat sink putty should be constructed near the weld area. With hammer welding, you form about 1/2in of weld, then set the torch down and quickly flatten the weld bead with a bumping hammer and body dolly before it has a chance to cool. The result can be a seam that is nearly as smooth as the original metal. If you achieve a seam of this quality, you are indeed doing professional work.

Where it is necessary to re-create the tack welds that were made at the factory, this can be done in two ways. One is to drill a series of holes in the patch at the approximate locations of the spot welds, then weld the two layers of metal together through the holes. The other is to drill holes through both layers of metal, then weld the holes closed. When the welds are ground smooth, they will look like the original factory spot welds.

Finishing Techniques

With the patch in place, attention is now directed to smoothing the seam. If you've used the

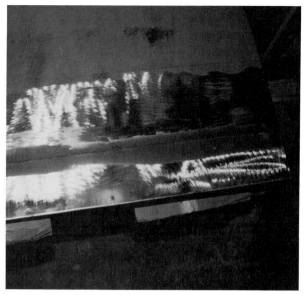

To make a butt joint, the patch is cut so that it fits precisely in place. Then it is tack welded in several locations. If the welder has enough skill and the joint is tight, the patch can be fitted so cleanly that the only finishing work is to grind the weld.

After the patch seams have been ground as smooth as possible, the repair can be finished with plastic filler.

hammer welding technique, the only finishing work required may be to grind the seam a little, then prime and fill remaining blemishes with Nitro-Stan. More likely, the seam will require some filler. Since no moisture should be able to penetrate the weld, plastic filler can be used. Or, if you'd rather, you can fill the seam with body solder—popularly called lead.

Plastic Filler

Plastic filler, better known as Bondo, has the association of sleazy back alley body repairs where holes and dents are filled with gobs of putty. The fact is, plastic filler is a thoroughly professional material and is used in top-quality metal repair.

Three guidelines lead to the successful use of plastic filler. First, since the material is water porous, it has to be applied over a solid metal backing. Second, only a very thin coating should be used; if a smooth surface can't be achieved with just a skim coating of filler, more metal finishing work is needed. If both of these guidelines are followed, the filler will easily last as long as the finish paint it supports. Third, purchase the filler from an automotive paint supply store. It's important that the filler be "fresh," and the quality of a product carrying the label of a major automotive paint manufacturer may differ from that of a "Brand X" product from a discount store.

To finish a repair with plastic filler, first make sure the metal is bare and clean. Then scoop a gob of filler out of the can and mix in hardener according to the directions. A piece of box cardboard works

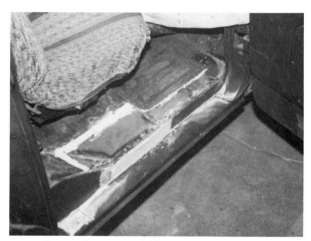

Note the small amount of filler used around the seams of this floor patch. When using plastic filler, the basic guideline is to apply as little as possible.

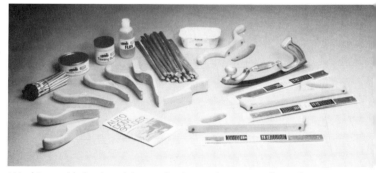

Working with body solder, or lead, requires a number of specialty tools and supplies. These include wooden paddles (used to spread the molten lead), body files, flux, tinning butter, tallow, and sticks of body solder.

well as a mixing palette. The hardener is worked into the filler until a uniform color change has occurred. Next spread a thin coating of filler over the repair area with a plastic applicator (available where you buy the filler). After the filler hardens—which takes about 20min—smooth the coating with a grater or a body solder file.

What you want is the least amount of filler required to smooth the surface irregularities of the repair. If a small amount of additional filler is needed for a smooth finish, mix up some more and spread it on the low spots. Grate or file this coating, and the repair area should be ready for priming and finish preparation. Any remaining small surface irregularities can be filled with Nitro-Stan.

Lead

Many hobbyists consider lead, as body solder is commonly called, to be the filler to use if metal repair is to be done "right." In fact, lead has risks and drawbacks. If solder flux gets trapped in the filler, it may work its way to the surface of the repair and blister the paint. When this occurs, the only remedy may be to remove the paint, melt out the filler, clean the metal, and reapply the filler. Working with lead also requires practice and some skill using a torch—neither of which are required for applying filler. Finally, the sticks of lead, flux, and related materials are more expensive than a can of plastic filler.

If you decide you want to boast that your truck contains no "plastic," you'll need the following supplies and equipment: several sticks of body solder (30 percent tin, 70 percent lead); tinning butter, as the flux is called; solder paddles; tallow, to keep the lead from sticking to the paddles; a soldering tip for your oxyacetylene torch; the torch; body files; and several cloth rags. The soldering supplies are available from restoration suppliers like The Eastwood Company.

The first step in applying body solder is to clean

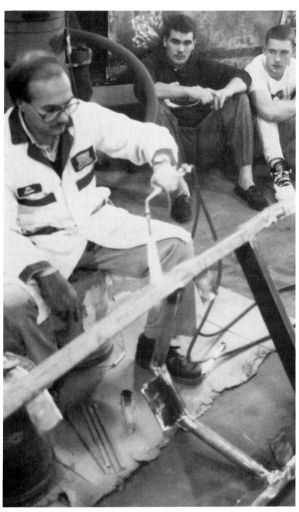

After applying flux to the area where you will be spreading the body solder, use a diffuse flame to heat the metal and melt the flux.

The key to working with body solder is to get the metal hot enough to soften the solder for spreading, but not so hot that the solder will melt and run off.

the repair area. Remove any paint with a dual-action sander, and clean the metal with Metal Prep or a comparable painting preparation product. Then gather the leading supplies and equipment; you'll want all items within easy reach. Now brush flux onto the area where you will be applying the body solder. Next slip the soldering tip over the tip currently on your oxyacetylene welding outfit and light the torch.

Work the flame around the repair area to heat the metal. The flux will begin to sizzle and boil. When the metal is hot enough, tin the repair area by melting a small amount of solder and spreading it over the metal with a rag. Wipe off any excess. Now reheat the area and melt a few chunks of lead so that they stick onto the body metal. Play the torch momentarily in the tallow tray, and dip a paddle in the heated tallow to coat the bottom of the paddle. Now work the torch over the repair area, melting the chunks of solder.

As the solder turns molten, spread it smoothly over the repair with the paddle. Unidirectional sweeping motions of the paddle work best. Keep the solder molten with the torch as you spread it with the paddle. As the tallow melts off the bottom of the paddle, you'll notice the paddle beginning to stick to the lead. The bottom of the paddle may also begin to burn. To keep this from happening, heat the tallow in the tallow pan and recoat the bottom of the paddle frequently as you spread the solder. When you've removed the heat from the metal while coating the paddle, you may have to play the torch over the solder again for a few seconds to make the solder liquid enough to be spread. When the solder has been spread over the repair area, you can shut off the torch and let the metal cool.

The last step is to smooth the filler with a body file. During this stage, it is wise to wear a painter's mask or similar breathing filter apparatus to keep from inhaling lead dust. If low spots show up in the

The solder is spread by dipping a wooden paddle in heated tallow and then working the paddle over the semimolten solder.

A torch is used to keep the solder molten as it is spread. The wooden paddle is recoated with tallow frequently during this process, to keep the solder from sticking and the paddle from burning.

127

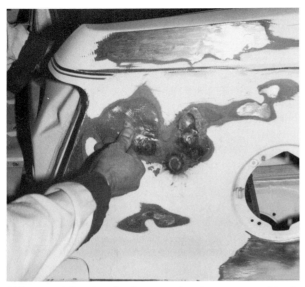

When straightening fenders or other large areas of body metal, it is possible to stretch and weaken the metal. When this happens, a condition called oilcanning occurs, where the panel becomes so weak that it can be popped back and forth with only a slight amount of pressure. The way to restore the metal's strength is through the metal-shrinking process described here. Metal shrinking is also used to repair warpage caused by excess heat during welding.

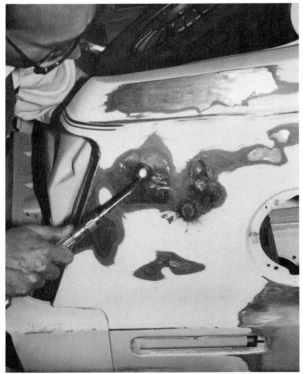

Metal shrinking can best be compared to putting little knots in the metal. The first step is to grind off the paint in the area where the metal has become weakened or warped. Next a dime-sized spot is heated with a torch until the metal glows red.

solder, these can be filled by heating the metal, melting a gob of solder, and spreading it over the spots with a paddle. As you file the solder, keep in mind that just because you have applied a metal filler doesn't mean the solder coating should be any thicker than if it were plastic. Once the solder has been filed smooth, wash the repair area with metal prep to neutralize any flux, then prime it and prepare it for finish painting.

The metal repair steps described here take a little practice—which can be done on scrap parts from a salvage yard—with patience making up for the lack of skill. You'll find that after you've repaired one rust or damage area, the next will be easier, and soon you'll be welding in patches and straightening dents with the touch of a professional.

Metal Shrinking

Whenever you work with metal, welding in patches or repairing dents, you risk stretching the metal. This stretching can occur in two ways. If you're bumping out a dent, it's possible to stretch and weaken the metal, causing what is called an "oilcan" effect. This condition occurs when dent work thins out the metal so that it loses its rigidity.

No trick is employed to tell if the metal has weakened enough for oilcanning to occur. Just press against the panel. If a section of the panel pops back with the ease that might cave in a soda pop can, the metal has been stretched and has lost its strength. When the metal is left in this weakened state, it flexes as you drive down the highway and someone pressing against the panel can easily dent the weakened section. Although you will be able to press the dent back out easily if you can reach the back side of the panel, the flexing metal will cause the paint and any filler to fleck or pop off, requiring more repair.

Metal stretching can also occur when heat is applied to the metal. The most likely occurrence is when you are welding in a patch panel. If the panel heats up too much, the metal may suddenly bulge. Or, as you finish the area you've been welding, you may discover a high spot that can't be flattened into the surrounding metal.

In either situation—stretching or heating—the solution is to shrink the problem area. This process can best be compared to putting little knots in the metal. Where the metal is weak, the knots provide strength to restore the lost rigidity. Where the metal has bulged, the knots shrink the bulge. Either heat shrinking or cold shrinking can be performed.

Heat Shrinking

Heat shrinking isn't an exotic art practiced only by skilled artisans. It is very easy to do. The most important ingredient is speed, so you will want to have the supplies and equipment you need laid out before you start. For equipment, you will need an acetylene torch with a welding tip, a torch lighter, goggles, a body hammer and dolly, and a small bucket of water and damp rag.

Begin by lighting the acetylene torch and adjusting the flame to the same neutral cone that you would use for welding. Next use the torch to heat a spot about the size of a dime in the flexible area or bulge until the spot glows red. Now work the metal around the heat spot with the hammer and dolly. The hammer needs to strike the metal at an angle directed toward drawing the metal into the heat spot. Although any smooth-faced body hammer can be used, special hammers with serrated surfaces that grab and pull the metal are made especially for heat shrinking. When the spot turns black, quench the heat with a damp rag. This combination of drawing the metal into the heat spot followed by rapid cooling tightens and shrinks the metal.

To strengthen a panel where the metal has been stretched so that it oilcans, several heat shrinks may be necessary. You can tell if the metal's strength has returned by pressing fairly lightly against the stretched area. If the metal still pushes in, more heat shrinks are needed.

If you are flattening a bulge, check to see if the excess metal has been absorbed by the heat spot, by placing a flat edge (the side of a body file works

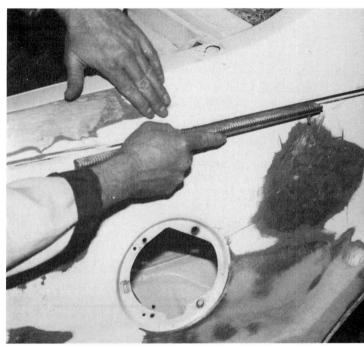

If you are flattening a bulge, you need to check if the excess metal has been absorbed by the heat shrink. This is done by placing a flat edge, such as the side of a body file, against the area where the warpage has occurred.

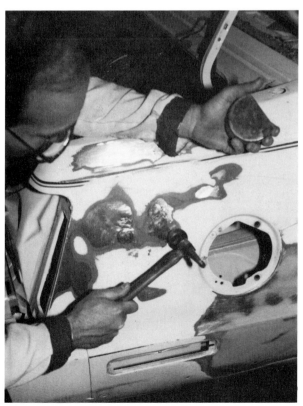

Using a hammer and dolly, work the metal around the heated spot in such a way as to draw excess metal into the heat spot.

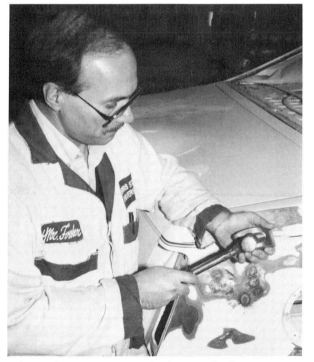

Vic Fowler, an instructor in the auto body program at Ferris State University, Big Rapids, Michigan, illustrates the tools used in cold shrinking. The grooved dolly is placed behind the warped or weakened area. A "tuck" in the metal is created by pounding with a ball-peen or rounded-tip body hammer.

Compare the fiberglass replacement fender, top, with the original steel fender, bottom. This quality replacement fiberglass part from Fairlane Company has mounting holes that match the originals, as well as original-style reinforcing brackets.

well) against the problem area. If the straightedge contacts the panel along its length—except, of course, on contoured surfaces—then the bulge has been removed. If some excess metal is still present, a few more heat shrinks may be needed.

Because heat shrinking crystallizes metal, it's not possible to work the repair area completely smooth with a body hammer and dolly alone—a process that's called metal finishing. In most cases, some filler will be needed.

It's best to practice metal shrinking before trying it in a real situation. For a practice panel, you can use a damaged fender from a bodyshop. To prepare the practice area, grind off the paint from a section 1ft or so square. To create the oilcan effect, just slap the section back and forth a few times with the heel of your hand until it collapses under thumb pressure. A bulge is easily created by heating the metal. The place to put the first heat shrink is in the middle of the stretched area or bulge. Several small shrinks—about dime sized—are better than a large shrink. Don't make the shrinks so that they touch each other; leave a little space between.

The fit of this replacement fiberglass fender installed on a mid-sixties Ford pickup is identical to that of an original steel fender. After the truck has been painted, the only difference will be that the fiberglass fender won't rust.

Heat shrinking is no more difficult or complicated than simple acetylene welding or metal bumping (it uses both skills) and is a necessary metalworking technique for anyone doing body repair.

Cold Shrinking

If you don't have a torch and need to shrink the metal on a body panel, either because hammer and dolly work has created an oilcan effect or to remove a bulge from earlier panel repair, you can cold shrink the metal. This is done with a ball-peen hammer and grooved dolly. If you don't have a grooved dolly, you can use a socket large enough for the hammer's peen head to fit into, or file or grind a groove into a flat dolly. What you are doing is putting a pinch in the metal. Cold shrinking leaves a "pinch" that needs to be covered with filler, but is otherwise as effective as heat shrinking for reducing a stretched area or flattening a bulge.

Fiberglass Replacement Body Parts

On trucks being restored to original standards for show competition, it is important to replace damaged body parts with new parts or original parts in excellent condition. But if driving your truck is your goal, then fiberglass replacement fenders, running boards, and hoods, available from various suppliers, may be a desirable alternative to expensive, hard-to-find metal parts. Fiberglass replacement body parts also have the advantage of being resistant to dents, immune to rust, and stronger than steel.

One manufacturer of fiberglass replacement body panels for vintage Ford trucks is Fairlane Company in St. Johns, Michigan. Fairlane produces top-quality fiberglass replacement fenders, hoods, and running boards, as well as custom "bubble skirts" for 1948–52 Ford F-1 trucks and 1953–56 and 1961–66 Ford F-100 trucks, that fit just like originals and look identical to the metal parts when installed.

Keith Ashley, Fairlane's owner, says the bottom line of his business is satisfying customers. The reason some fiberglass body parts have a reputation for being difficult to install on your truck is that steel fenders are flexible while fiberglass is rigid, so

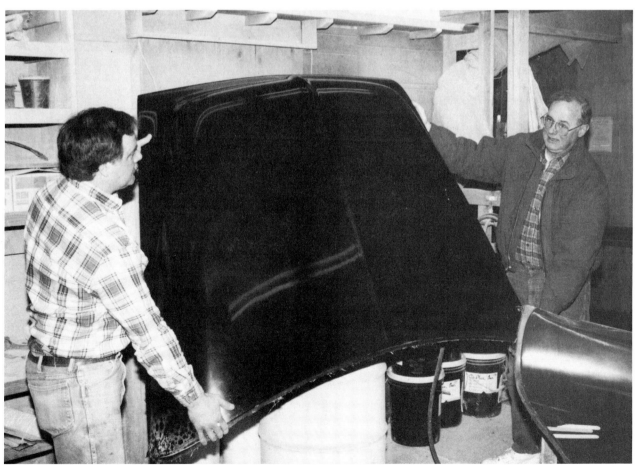

Fairlane Company owner Keith Ashley, right, and a helper remove a 1953–56 Ford F-100 hood from its mold. These fiberglass replacement hoods save restorers from the difficult metal straightening and rust repair needed on many original hoods.

The new fiberglass 1936 Ford pickup cab, fenders, and other body parts are molded to the same dimensions and contours as the originals, yet will fit any 1983 to present Ford Ranger longbed pickup chassis.

fiberglass body parts can't be bent or twisted in place. Steel, being flexible, can be manipulated into position. If the mold for a fiberglass replacement is taken from a new steel fender that has never been mounted on a truck, for example, it may contain distortion that the steel part acquired from years in storage. To make sure Fairlane's fiberglass replacement parts fit as they should, the steel original is first bolted onto a truck. A jig is built to hold the part in that shape, then the mold is made.

Also to make Fairlane fiberglass parts fit like the originals they replace, the replacement fenders, running boards, and hoods have bracings and mounting brackets molded in as needed. These reinforcing elements are made of fiberglass and steel, not the crude 2x4s sometimes seen in inferior parts. "If I can't make the part to be like the original, I won't make it," Ashley maintains.

Is the Ford in Your Future a 1936 Pickup?

Most of us are familiar with what are called replicars—cars with classic-era styling riding on a modern chassis and using late-model engines. Now a replitruck is available, a completely authentic-looking 1936 Ford pickup that has been molded in fiberglass by Coachworks of Yesteryear. Why re-create the 1936 Ford pickup, you say? Let's look at some reasons.

Anyone who has looked at the new-car sales charts knows that light trucks have incredible buyer appeal. And with the popularity of old-time styling, it makes sense that a new pickup wearing a classic 1936 Ford body would find a market. That's exactly what Coachworks of Yesteryear has done with its replitruck.

In explaining why he chose the 1936 Ford, Chuck Arnone, the replitruck's creator, said: "My love affair with the '36 Ford pickup goes way back to the mid-forties when my uncle Casper had one on his vegetable truck farm. It was so much fun to ride on the running boards and tailgate that I loaded lots of lettuce and celery in exchange for the pleasure of riding in that truck. Little did I know that one day I would replicate this neat truck."

Mid-thirties Fords—both cars and trucks—have a timeless styling that looks as clean and fresh today as it did new. But if you have a 1936 Ford pickup, chances are you're not going to drive it every day. The mechanical brakes are one reason. Not wanting to subject a fifty-five-year-old antique to the rigors of today's traffic is another. Neither of these concerns apply to a replitruck. Arnone has engineered the stock-dimensioned body to mate to any 1983 to present Ford Ranger 113.9in longbed chassis. If desired, the fiberglass pickup body and box can also be set up to fit either a stock Ford frame or a reproduction 1935–40 Ford frame of Arnone's design.

Combining the 1936 Ford pickup body and modern chassis gives a boulevard ride, interstate cruising speed, and the reliability of modern mechanical assemblies—with the look of a classic pickup. Coachworks of Yesteryear's new 1936 Ford pickup can be purchased either as a drive-away or in kit form. The kit—which includes the cab, hood, fenders, and complete box—has much of the hard work already done. The doors are prehung, saving all the hassle of alignment, and the bed is preassembled. These replitrucks aren't just 1936 Ford pickup look-alikes; the fiberglass body parts can be installed on original trucks, and the box even has Ford's famous script on the tailgate.

To keep the original look, the stock headlight has been replicated in steel and modified internally to accept either sealed-beam or halogen lights. Original-style Ford interior kits from LeBaron Bonney can be installed, or the truck can be fitted with a custom interior.

If you've been thinking about a replica project, this replitruck might just be it. Building a new vehicle is easier than doing a restoration, and quality replicas are holding their value. With the prices of classic pickups on the rise, it's logical to think that a well-constructed replitruck would be an inflation beater too. But most important, it would be a darn fun truck to own and drive.

The 1936 Ford is one handsome truck.

Chapter 11

Priming and Painting

Nearly every older pickup will need repainting. If your truck is undergoing restoration, its various parts and assemblies will go through the painting process at different times. If you are overhauling the truck, chances are you will repaint the entire vehicle at one time. Whichever approach you take, the steps in the painting process are essentially the same.

If you're like most owners of older Ford pickups, you'll want to find out what colors are authentic for your truck and make your color choice from this list. You may also be wondering whether or not your truck's current paint is an original color. If your truck has been repainted—and it very possibly has been—you can find what the original color was by checking locations that aren't likely to be repainted, like the underside of the hood firewall.

The colors of Ford pickups from 1940 to 1966 are listed in the appendices. You can get a sense of what these colors looked like from original Ford sales brochures, which can be purchased from antique car literature dealers at large flea markets. I say the brochures give a *sense* of what the colors looked like because in most cases, the colors shown on the brochures are likenesses, not exact matches.

You'll find that the original colors are still available. Unless you are dead set on painting your truck a nonoriginal color, it's better to stick with an original color scheme—and this includes the chassis. Not only do nonoriginal colors detract from your truck's value, but modern "wet look" paints that overzealous restorers often apply to chassis chip easily and are very difficult to touch up.

Before starting the painting process, you need to decide what type of paint will be used for the final finish. This decision has to be made now because the type of finish paint determines what type of primer will be used. Enamel finishes can be sprayed over lacquer or enamel primers, but a lacquer finish should not be applied over enamel primer—the reason being that lacquer solvents may penetrate and lift the primer coat. Modern urethane paints are designed as a system. This means that if you decide on a urethane finish, you should select compatible polyurethane base coats. With modern paints, particularly urethanes, selecting compatible primer-finish products is important for strong paint adhesion and to prevent the color coat from "sinking" into the primer.

Once the color and type have been selected, the next step is to decide where to buy the paint. If you are planning to use a modern paint—acrylic urethane or acrylic enamel—the local automotive paint store may be able to mix it for you. The other alternative is to order the paint from a vintage Ford truck supplier (see the listing in the Appendix).

Modern urethane paints are tougher than enamel and have nearly the gloss of lacquer finishes. Unlike the older paints, they are formulated as a system. This means the base coats must be compatible with the urethane finish. Since urethane paints are highly toxic, it is essential to observe the safety cautions associated with their use.

Primer Coat

The quality of the final finish lies in the care given to the primer coat. To obtain a glassy, smooth primer finish, you'll go through a multistep process that begins with making sure the surface to be painted is clean and rust free. If you are thinking about painting over an older finish, make sure rust isn't bleeding through from underneath and the paint isn't cracked. It's best to remove the old paint layers. If, however, the truck has its original factory coating and the paint is sound—dull and maybe worn through in places, but not cracked or blistered—it's OK to paint over it. But the old paint needs to be sanded so that it will allow a good bond with the new finish.

Primer Coat Preparation

To scuff up an old finish for repainting, you'll use a dual-action sander; a rotary sander would leave circular marks and gouges that would show up in the finish. Sanding an old finish by hand is not only extremely tedious and time-consuming, but will also result in an uneven base. A dual-action sander is an air-powered tool that operates in slow motion and is ideal for surface preparation. Before approaching the old finish with the dual-action sander, wipe down the paint with a wax remover solvent available at an automotive paint supply store.

If you have decided to strip the truck to a bare metal finish, you'll get better paint adhesion if you wipe down the fresh metal with dilute phosphoric acid, sold in automotive paint supply stores under trade names such as Metal Etch or Metal Prep. Instructions for applying the acid are given on the container. This treatment helps paint adhesion by turning the smooth metal into a microscopic landscape of hills and valleys. When the acid coating dries, it leaves a dull, grayish yellow coating, which also protects the bare metal from rusting. Untreated metal will quickly develop a rust coating if any humidity is present, even when stored inside.

Surfaces you don't want to paint need to be masked. Newspaper works well for this, but be sure to cover the tiny perforations at the edges of the sheets. Either overlap these edges or lay a strip of masking tape over the perforations; otherwise you'll find dots of primer on the underlying surface.

If your truck is still assembled, you will need to unbolt the box and slide it back or lift it off the frame to prime the rear of the cab and front of the box.

Primer Coat Products

Years ago, it was common to use a single primer product. Today several specialized primers are used at various stages of the surface preparation process and for special applications. Besides a basic primer, which serves as a base coating for subsequent primer and finish coats, primers that resist corrosion are also available. These are used on inner body panels and chassis parts where moisture can cause rust problems. Examples of corrosion-resistant primers are zinc chromate, available from most painting suppliers, and The Eastwood Company's Cold Galvanizing Compound. These primers are not intended to be sanded.

The next surface preparation product is primer-surfacer, which is typically applied over the primer layer. Primer-surfacers build up more quickly than basic primers and can be sanded to fill minor surface irregularities. Rough surface areas where rust has been removed by sandblasting will require an extrathick primer buildup that can best be achieved by a sprayable filler like Sandy, available from The Eastwood Company.

When a perfectly smooth primer surface has

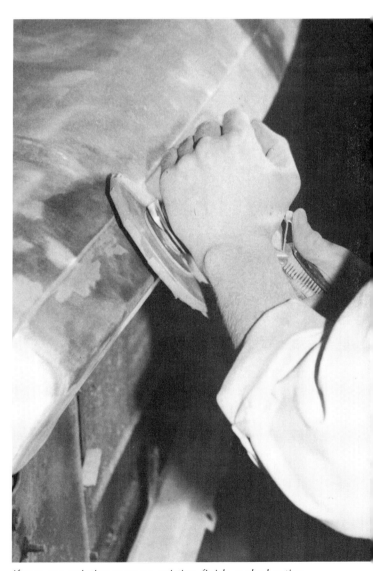

If you are painting over an existing finish, a dual-action sander is used to scuff up the old paint to ensure good adhesion of the new base coat. A dual-action sander can also be used to remove the old finish.

If the old finish doesn't have a lot of scratches and cracks, it can be used as a base for the new finish. An important preparation step for repainting is to remove all trim. Note that this includes the bumpers and taillights as well as the door handles.

The primer coat serves two purposes: to bond the finish with the metal and to provide a base for the finish coat. The final finish will only be as smooth as the primer base, so several primer coats are usually applied.

All dent and rust repair is done before painting. On this truck, the lower areas of the doors and box, where surface rust develops, have been stripped to bare metal. All bare metal areas should be treated with dilute phosphoric acid to remove any traces of rust and to etch the metal for a good paint bond.

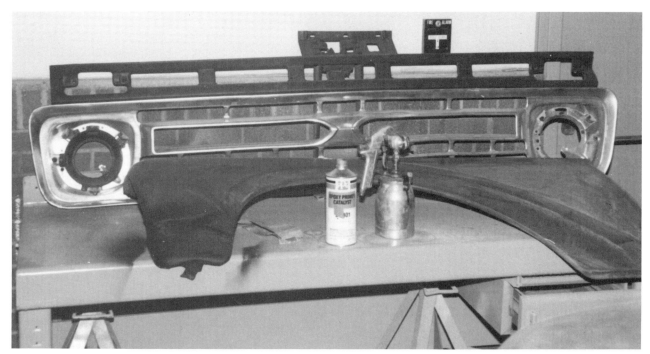

If you are working on a show restoration, the fenders and doors will often be primed and painted while off the truck, then carefully reassembled. This approach allows all surfaces to be painted.

been achieved, a primer-sealer is applied. The sealer prevents sand scratches from showing through the final finish and gives the color greater depth.

Primer Coat Application

All primers are sprayed with a pressurized paint-spraying gun. Lacquer and enamel primers are diluted with solvent. Catalyzed primers—so-called because a "hardening ingredient" must be added in order for the paint to dry—are mixed with an equal proportion of catalyst.

Lacquer primer is diluted 125 to 150 percent with thinner solvent; this means you will add somewhat more than an equal amount of thinner to the primer. Enamel primer is mixed with a 33 percent volume of reducer solvent; this means you will add an amount of reducer equal to one-third the volume of enamel primer. Thinners and reducers are sold in various grades and types. Primers do not require as high a grade of thinner or reducer as do finish paints. Thinners and reducers also vary in their evaporation properties. If you are priming or finish painting in a warm temperature (above 75 degrees Fahrenheit), you will need a solvent that is less volatile, or slower to evaporate, than if you are spray painting at a cooler temperature (60 to 65 degrees Fahrenheit). The counter clerk at an automotive painting supply store can tell you the correct solvent for the type of paint and temperature conditions.

The mixing proportions of primer and solvent or catalyst are very important. A good way to make sure the mix is correct is to cut the top of an empty

---

Priming and Painting Products and Steps

The following list describes the type of finish you will give to each surface in the priming and painting process. It starts with the first finish to be used after you have stripped and cleaned the metal, and continues in order of application.

1. Over bare metal,

    a. treat the metal with dilute phosphoric acid (Metal Prep or a comparable product) to retard rust and improve paint adhesion, and

    b. coat areas prone to rusting (the inside of doors, the lower areas of the cab) with a zinc-rich primer such as zinc chromate.

2. Over bare metal or a prior finish,

    a. use epoxy or traditional lacquer or enamel primer (gray, red oxide, or black), and

    b. apply a dust coat of a contrasting color for block sanding.

3. Over a base primer coat, to fill surface imperfections, spray on a high-build primer-surfacer.

4. Over a primer coat, as preparation for the finish coat, use a primer-sealer or sealer.

5. Over a sealed primer coat, apply a finish paint in acrylic lacquer or enamel or modern catalyzed acrylic urethane.

6. Over a finish coat, consider spraying on a clear top coat of acrylic urethane for extra-high gloss and a "wet" look (this step is required with some urethane system paints).

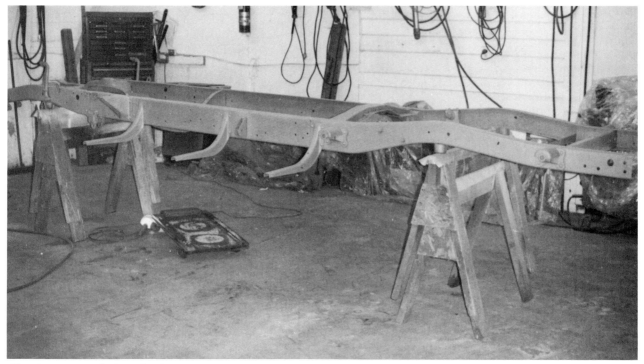

Chassis parts, including the frame, often receive a zinc chromate primer base. This coating helps prevent rusting.

To paint the back of the cab, as well as to clean and finish the frame, it is necessary to remove the pickup box.

solvent can; pour in the primer; and then add the solvent or catalyst—whichever the product requires. A graded measuring stick, available from the painting supply store, will show when the correct mixture has been reached. Now the mix needs to be stirred.

When filling the spray gun cup, always pour the mix through a paint strainer. Cone-shaped paper strainers specially made for this purpose are available from automotive paint supply stores.

Even if you don't have a lot of spray painting experience, you don't have to be afraid of putting on the primer. The air line pressure should be 35 to 45psi for lacquer, 55 to 60psi for enamel, and 40 to 50psi for catalyzed primers. The temperature should be between 70 and 85 degrees Fahrenheit. The spray gun is held a distance of 8 to 10in from the painting surface. If these guidelines are followed, virtually anybody can apply the primer coating.

With the primer coating, it may be desirable to build up a thicker layer in repair areas. This is done by applying several coats, allowing drying time between each. When paint runs occur, the only loss is the time required to sand them out and reprime for a smooth surface. The main intent is just getting the primer onto the vehicle.

Primer Coat Sanding and Smoothing

Before sanding, the primer needs to cure. Times and temperatures for curing will be listed on the container. If you try to sand fresh primer before it has dried sufficiently, it'll probably roll up in little balls on the sandpaper.

Traditional lacquer and enamel primers should be dry sanded. This is a dusty job, so you'll want to wear a dust mask or painting respirator. Some primers can be wet sanded. This is a technique where automotive-type wet-dry sandpaper is dipped in water periodically while you are sanding. Wet sanding keeps down the dust and extends the life of the sandpaper.

Ask the counter clerk at the automotive paint store where you buy the primer if the type you're using can be wet sanded. Most of the catalyzed primers seal out moisture—which means they can be wet sanded. Most of the older-style lacquer and enamel primers, however, are water porous, which means that wet sanding would allow water to penetrate to the metal surface—which can cause rust and may eventually blister and lift the finish paint. For this reason, also, primed truck bodies or parts metal should not be stored outside where dew or rain can settle on the primer surface.

To wet sand, use automotive sandpaper with a water-resistant backing and place a bucket of water near where you are sanding. Periodically dip the sandpaper in the water to wash off the sanding dust. It also helps to moisten the primer surface to "lubricate" the sandpaper. This can be done by washing the primed area with a wet rag or sponge, or by filling an empty Windex or dish detergent bottle with water and squirting it onto the sanding area.

The life of a sheet of automotive sandpaper is extended by folding the fresh sheet in half in the long direction, then ripping the paper at the fold. Now fold the half sheet in thirds, as you would fold a business letter. By doing this, you have created three sanding surfaces. If you are dry sanding, you can tap the sandpaper against a board or other hard object periodically to shake off the sanding dust. When one surface becomes clogged, refold the sandpaper and continue sanding with a fresh surface. This way, all working areas of the sandpaper get used.

The best way to avoid a ripply surface is to clamp the sandpaper in a sanding block instead of holding it in your hand. When you hold the sandpaper in your hand, the parts of the paper that are under the base of the knuckles and palm cut deeper, causing an uneven surface.

Coarser-grit sandpaper is used for sanding filler coats, and finer grit for the coating that will form the base for the color coat; sealer coats are not sanded. A coarse, 220-grit paper works well for smoothing the filler coats but leaves scratch marks that need to be filled by additional primer coatings, which are sanded with progressively finer, 360-, 400-, and 600-grit sandpaper. The final primer coat is then sealed.

The primer coat is carefully wet sanded to ensure a smooth base for the final finish. Here the bodyman is using a dish soap container filled with water to wet the primer, while carefully sanding out small surface imperfections.

Powder Painting

For smaller parts, particularly, powder painting can be a desirable alternative to spray painting. Actually the coating used is not paint, but a dry, granular material. Because the product cannot be tinted, the color of the powder is the color you get. The selection of colors is broad, however, so unless you are trying to get an exact color match for authenticity's sake, you can usually find a color that is "close enough" to use on your truck.

Parts being powder painted must first be sandblasted or chemically stripped to bare metal. They are then hung on a metal frame in a large, enclosed cabinet. The parts are given a negative charge, and the powder is sprayed into the cabinet, where it is electrically attracted to the charged metal. The coating is then baked for several hours at around 500 degrees Fahrenheit. A finish much tougher than paint results.

Because of the electrical bonding and baking processes, powder paint can't be applied over filler. It is also a fairly expensive finish. Examples of parts you might want to have powder painted are emergency brake levers, sideview mirror brackets, and other small parts of this nature.

Filler putty can be used to repair small surface nicks or scratches. Only a very thin coating of filler is applied.

One hard part of the priming process is knowing when the primer surface is smooth. Since primer is dull, it is difficult to see sand scratches, dips, and waves that will show through in the finish coat. One way to make sure the surface is smooth is to spray on a light coating of contrasting color primer—gray primer over red, or vice versa, for example. As this guide coating is sanded away, spots of contrasting color will show high or low areas that needed to be sanded more or filled.

To smooth out nicks, deep sand scratches, and other surface imperfections, filler putties are used. The filler putty applied with lacquer and enamel primers is called Nitro-Stan. The corresponding product used with catalyzed epoxy primers is called Poly Putty. Nitro-Stan comes in a tube and is applied by squeezing a small amount onto a rubber applicator and then wiping it onto the nick or scratch. Poly Putty comes in a metal container and is mixed with hardener. Only a very thin coat of either filler is applied. When dry, the filler can be sanded using the block sanding technique.

Sealer Coat

The finish coat can be sprayed directly over primer—provided compatible primer and finish paints have been selected—but for best results, a sealer coating should be applied before the final

---

Precautions to Take When Using Isocyanates
by Alvin Shier

Many hobbyists use automotive paint products without realizing the potential health risks involved, or their seriousness. The highest-risk products are those containing isocyanates.

Just what are isocyanates? Simply, we could say that any automotive product that requires an additive to catalyze, or "kick over," the material contains isocyanates. Some of the more common trade names are Delthane, Endura, Indural, Multrathane, and Vibrathane. Isocyanates are found in all plastic filler and catalyzed paint products.

In their solid and liquid forms, isocyanates can cause an irritation or burn on contact with skin. They cause irritation of the digestive tract when ingested. This can lead to severe permanent damage. No medical cure is available. Prolonged ingestion in the vapor form can cause a sensitization that affects the lungs, and thus the breathing. Lung sensitization is the most dangerous health hazard associated with isocyanates.

With all the potential problems, how can we paint at the hobby level and minimize the health risks? To be 100 percent safe, we must keep clear of any catalyzed product such as plastic fillers and acrylic urethane paints. With some common sense, however, we can work at home with these products by using the following guidelines:

1. Always read and understand all precautions for the product being used.

2. Work in an area with the most ventilation possible—outside is best.

3. Do not allow any hardeners or catalyst to come in contact with clothing or skin.

4. Work fully clothed. Isocyanates in their vapor form penetrate the skin and enter the blood stream.

5. The minimum protective requirement is the best charcoal respirator you can buy. The minimum requirement in many U.S. states and Canadian provinces is an external oxygen-fed mask.

6. Store any remaining product out of the reach of children or other people who may not understand the problems associated with its use.

If you have any doubts about the paint products or their chemical composition and you are allergy prone or have respiratory ailments, please consult your physician before proceeding. It could turn out that the least expensive way for you to paint will be to save enough money to have the job done by a professional in a clean and safe environment.

When the primer coat is smooth, the next step is spraying on a sealer coating, which is followed by the final finish.

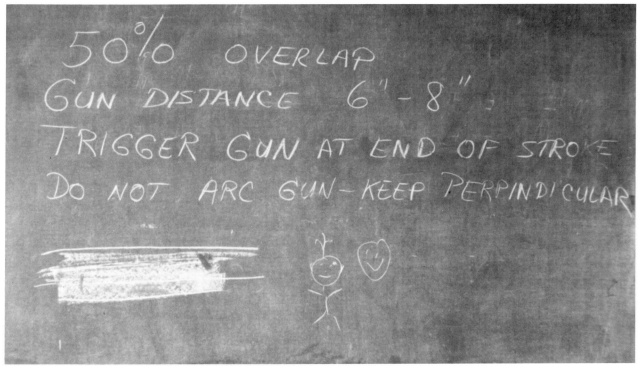

Instructions for finish painting are shown here. Each pass should overlap by one-third, the paint gun should be kept 8in to 10in from the painting surface, and the gun should be held perpendicular to the area being painted—not swept in an arc.

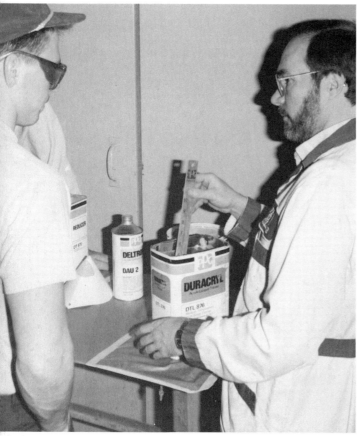

Mixing the correct proportions of paint and solvent is very important. A graded measuring stick (available from a painting supply store) will show when the correct combination has been reached.

With a truck undergoing a frame-up restoration, each component is finish painted, then assembled—beginning with the frame.

Finish painting follows a sequence that starts at the top of the truck and works its way around the body. Here the top has not been painted because it will be given a contrasting color.

finish. Sealer is available as a clear coat that is not sanded and serves only to form a moisture and solvent barrier, or as a combination primer-sealer that can be sanded but forms a hard moisture and solvent barrier as it cures.

The only drawback to using any sealer is that if you wait an extended period of time to do the finish painting, the sealer becomes very hard and makes a poor bond with the finish coat. When the sealer hardens, it also becomes very difficult to sand out any scratches that can occur from carelessness.

Finish Coat Preparation

It is very important that the surface be completely free of dust before painting. To remove sanding dust, first use an air nozzle to blow off the areas to be painted. Be sure to blow into seams and crevices on the vehicle, as well as into seams and folds on the masking paper, where dust might be trapped. Then wipe down the entire surface with tack cloths (available at automotive paint supply stores).

As mentioned earlier, if you are doing a ground-up restoration, the painting process will occur in stages. Typically the chassis is painted first, then the cab interior, followed by the inside of the fenders and hood, the cab assembly with front fenders and

hood attached, and finally the box. With this approach, unless you are careful, preparing the next assembly for painting can mess up the parts already refinished.

One way to keep from spoiling a freshly painted chassis while you are preparing the cab assembly is to cover the chassis with a large sheet of plastic (available from building and farm supply stores), then mount the cab and front end assembly over the plastic. Now you can prime, sand, and paint the truck's cab and forward sheet metal while keeping overspray, sanding dust, and wet sanding drippings from mucking up the chassis. When the painting process is completed, you can tear off the plastic and everything underneath will be spotlessly clean.

If you're painting an assembled truck, cover the engine to prevent sanding dust and painting overspray from settling on the engine and its associated wiring and electrical parts. You can protect tires from paint overspray by covering them with garbage bags or jacking up the truck, supporting it on jack stands, and removing the

---

Painting Tips for Ford Trucks

Cast-iron parts can be painted with a "cast-blast" coating, which gives the look of fresh cast metal.

To paint the letters on the hood nameplates, rather than trying to mask the individual letters, paint the lettering and compound the excess off the plate, using a piece of leather to work the compound.

When painting the engine, don't put too much paint on the block. Paint reduces heat transfer.

Color the top primer layer for the undercarriage black, and chips in the final finish won't show a drastic color change.

When painting trim and other smaller parts, hang the pieces from wire strung between stepladders.

When fitting painted parts during reassembly, tape off the edges with 2in masking tape, then pull off the tape after the parts have been fitted, just before bolting them up.

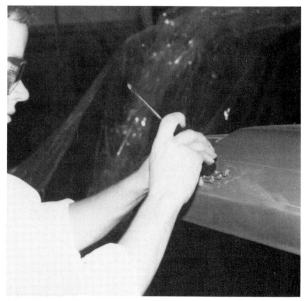

After the finish coat has been allowed time to cure and harden, the exterior trim is replaced.

After the finish coat has dried, it's a good idea to cover the truck with light plastic sheathing to keep dust off the finish and to prevent accidental mars or scratches.

As one of the last steps, the Ford letters on the tailgate are given a contrasting coating.

wheels. The new finish will look much more professional if you take the time to remove nameplates, model markings, and other trim.

It is important to buy enough finish paint to do the whole truck. When mixing the paint with solvent, it's a good idea to add paint from different cans. This prevents subtle—or sometimes quite visible—changes in the color caused by slight variations in the mix. Even with paint of consistent color, shading variations can appear in areas that are applied on different days under different temperature and humidity conditions. The recommended approach is to spray all areas of one color at one time.

Finish Coat Application

For the less-experienced painter, faster-drying lacquer paints are easier to apply than slower-drying enamel. With slower-drying paints, runs are more likely to develop and it is harder to get that "wet" look on the finish coat that is important for a high gloss. Modern acrylic enamels combine the fast-drying characteristics of lacquer and the toughness of enamel, so they make a good paint for the hobbyist restorer. The drawback to acrylic enamel is a somewhat subdued gloss.

The other option is to apply a urethane paint. Urethane finishes would seem to have all advantages and no disadvantages. They have a high gloss and a tough finish, dry rapidly, and are relatively easy to apply. These modern high-tech paints are extremely toxic, however, and require special respiratory equipment to prevent serious illness, possibly even death. Precautions printed on the cans of today's painting products are not to be taken lightly.

Whereas primer can be sprayed in the open shop, finish painting requires dust-free conditions—preferably a painting booth. Just because your garage or shop doesn't have this expensive facility doesn't mean you have to hire out the finish painting. You can jerry rig a spray booth by sectioning off a portion of the shop with large sheets of plastic. You will need to install an exhaust fan in a window or in an opening cut into a wall, to draw painting overspray and solvent fumes out of the spraying area. You will also need to have a ventilation inlet for fresh air. The ventilation opening, which can be windows or holes cut in the plastic, should filter the incoming air. Furnace filters work well for this.

It is also important to wear a professional-quality respirator. You'll get better protection against toxic chemicals from those with a charcoal-activated filter than from the type with paper filters. If you're working with urethane paints, the respirator to use is the type that draws the air supply from outside the painting area. This fresh air-style respirator is described later in this chapter, under "High-tech Respirators" in the "High-tech Painting Equipment" section. Modern painting products containing isocyanates are very dangerous if sprayed without adequate ventilation and air filtration.

Think out the pattern you're going to follow, before starting to spray the final finish. The typical pattern is to start with the upper portion of the cab, then work around the truck from front to back or back to front. Before spraying any paint, be sure the air vent on the paint cup lid is opposite the spray nozzle. Otherwise paint will drip out of the gun onto the hood or cab roof when you hold the gun over these horizontal surfaces.

The common spraying approach used by most inexperienced painters is to sweep the spray gun in an arc across the painting area. An arc motion puts more paint in the center of the sweep than at the starting and ending points. The paint layer must be of an even thickness, so the correct approach is to keep the painting nozzle pointed directly at the painting area and hold the gun an even distance—8 to 10in—

Wheel Pinstriping
by Roy Nagel

I remember watching a very talented bodyman—a man who worked for my late father and great-uncle in the paint shop at a Studebaker dealership they owned back in the late forties—apply wheel stripes a number of times.

He would begin the process by getting one of the hydraulic "roll-a-car" jacks from wherever they were in the shop. (I'm referring to the kind of jack that has a round horizontal lifting pad that is about 5in in diameter and can rotate in respect to the jack frame, two small wheels in front and casters at the back, and an approximately 4ft handle that one pumps up and down to raise a vehicle.) Then he would pump the jack up to its maximum height and set one of the already painted wheels—usually with a mounted tire—on the raised jack pad.

He then took a striping brush in hand, dipped it in whatever color he wanted, took a few practice strokes on whatever scrap panel was handy—to get just the right amount of paint on the bristles—and touched it to the wheel surface. Using his knee to steady his right arm as he crouched beside the tire, he rotated the tire and wheel with his left hand to make a complete circle. He'd repeat the process to get as many concentric stripes as one wanted.

I'll bet it took him a lot of practice, but he sure made it look like a snap! On the few occasions where I saw him wind up with a stripe that varied in width or didn't quite match up, he would wipe the wheel off with a rag and do the whole thing again from scratch.

I don't know what kind of paint he used, but I can't recall that he ever left a newly striped wheel sitting stationary more than a few minutes before taking it off the jack surface to start on the next one.

from the truck.

In painting a panel, make one complete sweep, then move the spray gun down the panel so that the next sweep will overlap the painted area by about one-third. For this next sweep, bring the gun across the panel in the opposite direction. Now continue overlapping the bottom third of each previous sweep and changing the direction that you move the gun, until the entire panel is painted. Then move to the next panel and paint it following the same approach.

You want the finish coat to have a glossy sheen and a smooth surface, and the secret to a supergloss finish is nothing more than several coats of evenly applied paint. Then when the paint has cured, it is microsanded using very fine, 1,000- to 1,500-grit sandpaper. Microsanding removes the texture from the paint, leaving a very smooth but dull finish. To bring out the gloss, the finish needs to be compounded. The end result will be a mirror-smooth paint job that glistens like a diamond.

To build up a finish coating that is thick enough to be microsanded and compounded, you'll make one complete pass around the truck. Usually by the time you are back where you started painting, the first coat will have started to set up (become tacky). If so, you can continue spraying the next coating. If the paint is still fresh—which is the way it's likely to be if you are using a slower-drying enamel or are painting in colder or more humid conditions—you'll need to wait awhile until the finish has a tacky, or sticky, feel. You can test the condition of the paint by touching a finger to overspray that has landed on some masking paper. If you start with the next coating while the paint is still wet, you'll soon see the finish sag and drip. If that happens, wait until the paint becomes tacky and then continue spraying. The runs can be sanded out after the finish dries. You'll probably need to prime the area where you're fixing the runs, and then you can repaint it.

Typically four coats—four passes around the vehicle—are needed to get enough buildup for microsanding. Make sure the lower panels also get

The final step is compounding the finish to bring out its luster. This is best done with a power polisher.

full paint coverage. This will require stooping down when you paint the bottom of doors and fenders.

Before it is microsanded, the finish needs to cure. Instructions on the paint container will give the recommended curing time.

Microsanding is done using a block sanding method. Be very careful to stay away from sharp curves or edges in the body panels. These edges have less paint buildup, and if you sand them, you will expose the primer very quickly. To keep from accidentally sanding into an area with little paint buildup, place masking tape over all sharp edges and curves. After you have completed the microsanding and have machine compounded the finish, you will carefully hand compound these areas.

After it has been microsanded, the finish will have a dull, flat look. The gloss is restored by buffing the finish with polishing compound. Hand buffing a truck's finish is very hard work and time-consuming, so the compounding should be done using a power buffer. As described earlier, to keep the buffer from cutting through the paint on styling creases, first mask any sharp bends in the metal.

Clear Coat

Those who are preparing their trucks for the show circuit often cover the final color layer with a clear coating. The clear coat protects the finish and enriches the color by giving it the appearance of having more depth. A clear coat also eliminates the need for frequent waxing to prevent the paint from oxidizing. The drawback to clear coating is that spot repairs—to touch up stone chips or other damage to the finish—are difficult to make.

The two primary ingredients to a quality finish are thorough preparation, and microsanding and compounding the final finish. If you use compatible painting products, apply them in the correct sequence, follow the mixing formulas on the cans, and do each step according to the guidelines given above, you'll feel a swelling of pride when admirers ask for the name of the shop that painted your truck.

High-tech Painting Equipment

Today's two-part paint products—the type where the paint or primer is mixed with a catalyst—are extremely toxic, to the point of being life threatening if used carelessly. The ability to purchase these products conveniently from any automotive painting supply store with no cautionary warnings other than those on the product label, presents a real danger, given the hobbyist's easygoing attitude about personal safety. So let's wake up and realize that we shouldn't use these high-tech paint products unless we understand and follow the safety and environmental precautions that they demand.

High-tech Respirators

The days of spray painting without a proper respirator are gone. No automotive painting product made can be sprayed safely wearing only a disposable dust mask as protection against painting fumes. At the minimum, you should use a professional-grade painting respirator. If you are spraying the high-tech paints with their deadly isocyanates, the type of respirator to use is one that supplies air from a source outside the painting area. With this style respirator, the face mask draws air through a hose that is attached to a compressor located in an area with fresh, uncontaminated air. Although the air hose may seem cumbersome and restrictive at first, knowing that the air you're breathing is free of toxic paint fumes is worth the small inconvenience.

Modern paints are not to be fooled with. To ignore the toxic chemicals in today's high-tech paints is to invite a condition known as acute respiratory distress syndrome, which closely resembles a heart attack. If the expense of the necessary safety equipment is outside your budget, either avoid using the modern two-part painting products or have your truck painted by a professional.

The days of spray painting without a proper respirator are gone. If you are spraying the high-tech urethane paints with their deadly isocyanates, the type of respirator shown here, which supplies air from a source outside the painting area, should be used. The Eastwood Company

High-tech Spraying Equipment

Along with the paints, paint-spraying technology is also changing. Traditional paint-spraying equipment uses high pressure—35 to 80psi—to atomize the paint and blast it onto the vehicle. High-pressure paint spraying produces a fine-quality finish, but 70 to 75 percent of the paint and solvent blows off the vehicle—either landing on some other surface or being exhausted into the atmosphere. (If you're not wearing a proper respirator, you're breathing in some of this overspray.) Not only does this represent a waste of expensive paint, but the overspray that diffuses into the air is doing serious damage to the atmosphere. As a result, clean air legislation in several states is starting to restrict the use of automotive painting solvents and is urging that less wasteful technologies be used in applying automotive finishes.

A much less wasteful spray painting technology, called HVLP (High Volume, Low Pressure), practically eliminates overspray; 90 percent of the paint stays on the vehicle. Although an HVLP paint-spraying system costs more than a good-quality painting gun purchased alone (the purchase of most HVLP equipment includes the compressor), the savings in paint and solvent, a reduced health risk, and environmental benefits—plus the likelihood that this technology will be mandated in the near future—makes HVLP paint-spraying equipment an investment definitely to consider.

What makes an HVLP painting system different is the turbine compressor that puts out a very high volume of air. Unlike the average hobbyist's portable air compressors that may have an air supply capacity of 7 or 8cfm, the HVLP turbine compressor will deliver from 60 to 90cfm of air. Since air becomes hotter as it is compressed, and because heat dries up moisture, the air delivered to an HVLP painting system is both warm and dry—ideal for automotive

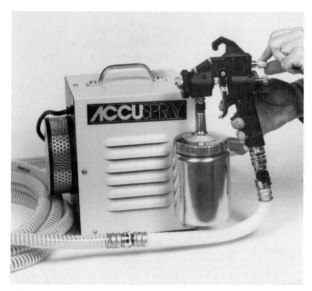

Along with paint products, paint spraying technology is changing. A much more efficient technology called HVLP uses a turbine compressor to feed between 60cfm and 90cfm of air into the spray gun. This large air supply allows the paint to be applied to the vehicle at very low pressure, greatly reducing overspray and toxic fumes. The Eastwood Company

The HVLP compressor sits outside the spraying area.

Engine accessories can be given a factory-fresh appearance with specialty paints like Eastwood's Detail Gray. The Eastwood Company

painting.

Very little training or, for experienced painters, reeducation, is needed to operate an HVLP system. An HVLP spray gun is handled in exactly the same way as a traditional paint-spraying gun. The only difference is in what happens to the paint after it leaves the gun. With the HVLP system, the paint is sprayed onto the vehicle surface with just enough force to adhere and flow. Almost no overspray results.

If you are wondering which type of spraying equipment to buy to paint your truck, keep in mind that you'll have no difficulty learning how to use an HVLP system. If you have spray painting experience, switching to an HVLP system won't be any harder than getting used to a new spray gun.

Because the technology is new, HVLP systems have some disadvantages. Metallic paints don't spray well, and the finish is sometimes rougher than would be the case with high-pressure painting. If you're using a standard, or nonmetallic, paint and will be microsanding and compounding the final finish, neither of these problems should be a cause for concern.

HVLP spray painting is more economical, in terms of materials costs, and safer to the painter and the environment. These reasons recommend it to the hobbyist's as well as professional's use.

Correct Paints

Black is black, right? Wrong! The correct chassis black is neither glossy nor dull. Called Eastwood Chassis Black/R #1244, this paint has a specially formulated epoxy base and high solids content, to give four times the coverage per can. Three cans will paint a typical chassis.

Often steel parts had no finish from the factory, but to control corrosion and maintain the high level of your restoration work, you should apply some protective coating. Eastwood has developed the correct paint to protect your exposed steel parts while retaining a steel appearance. This paint, Detail Gray #1246/R, has been formulated to give not only the proper shade, but also the proper surface texture to raw steel parts. Order from the Eastwood Company (contact information listed in Appendix).

Not all blacks are black. This paint matches the chassis black applied at the factory. The Eastwood Company

Chapter 12

Replacing Wiring

If you're wondering whether or not your truck's wiring needs replacing, the simplest way to make that decision is to take a close look at the condition of the wiring. If the insulation is cracked or maybe even missing in places, or new wiring has been spliced in and old wires wrapped with electrical tape, a new harness is definitely in order. Failure to replace deteriorated wiring invites problems on the road, burned-out electrical equipment, or an electrical fire that can destroy the truck as well as the building in which it is stored. If you have any doubt about your truck's wiring, replace it. As you will discover in this chapter, the procedure is not difficult and does not require electrical expertise. On a frame-up restoration, you will install a new wiring harness as a matter of course.

Tools and Supplies

To install the new wiring harness, you will need a shop manual with a wiring diagram; pliers; a wire

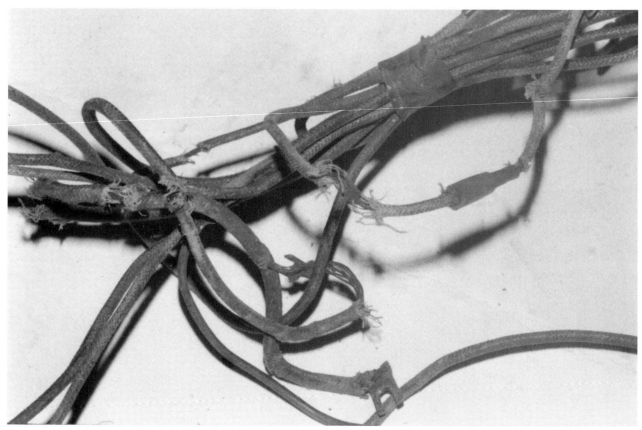

If you're wondering whether your truck's wiring needs replacing, just take a close look at it. If the insulation is cracked or missing, new wiring has been spliced in, and old wires are wrapped with electrical tape, then a new harness is in order.

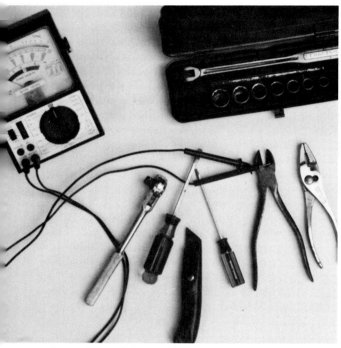

The tools needed to replace a wiring harness include a 3/4in socket set, screwdrivers, cutter and regular pliers, a utility knife, and a test device.

stripper; butt connectors; assorted screwdrivers, regular and Phillips; a socket set; ViseGrips; a test light—a VOM is also desirable; and in some cases, a steering wheel puller.

Place your order for a new harness several weeks before the time you plan to install the new wiring. Order the harness only from a reputable vintage Ford parts dealer or harness manufacturer. "Bargain" harnesses from discount mail-order outfits are often so far from original standards in color coding, wire size, and wire length that they either have to be remade to fit the truck or have to be discarded and replaced with a quality version from a supplier such as Rhode Island Wiring Service.

A quality wiring harness will be constructed of the correct gauge wire and covering and will follow the original color coding. The correct wire gauge is important because older 6-volt trucks require larger-diameter wire. To be authentic, wiring harnesses in trucks built before the mid-fifties should have fabric-covered wiring. Without correct color coding, the harness will be almost impossible to install.

Preparation

With a quality harness, you should receive an instruction sheet showing hookup connections. For a frame-up restoration, the wiring is removed when the truck is disassembled. This means that you will need to refer to photos you took of your truck before assembly, as a guide to where to route the harness. If you forgot to take the photos or the wiring was

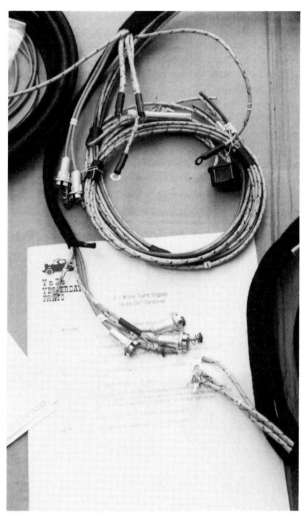

A quality harness will use original-style insulation, the wires will be color coded to match your truck's wiring diagram, and the end clips and light sockets will be correct for your vehicle.

already out of the truck when you did, you can take photos or make diagrams showing the location of the wiring in other original or correctly restored trucks.

As a double check of the instruction sheet, you should also have a service manual with a wiring diagram for your year and model truck, to refer to as you install the harness; wiring diagrams can also be purchased separately from a service manual. The wiring diagram will help you understand your truck's electrical system and may be needed later to troubleshoot any problem circuits.

Before replacing the wiring, it's a good idea to spend enough time studying the wiring diagram so that you understand the major circuits and are familiar with the color-coding scheme. The wiring diagram shows circuits by labeling each wire with its color, and sometimes its gauge as well. As you become familiar with the wiring diagram, you will be able to trace various color wires from a component—the taillight, for example—to their power source or ground. Being able to read the wiring diagram will be a great help if you become confused, as you are installing the harness, over which wire attaches to which terminal on a component; you will be able to quickly find the answer in the wiring diagram. As you study the wiring diagram, you will also realize why it is important to have a correctly color-coded harness.

Wiring diagrams for trucks built before the mid-fifties won't show signal lights because before this time, signal lights were an aftermarket add-on, installed by either the dealer or the owner. These aftermarket signal lights were usually mounted on the top of the front fenders and at the corners of the box, and were turned on and off by a lever that clamped to the steering column. Few vintage truck owners like their look, but signal lights are important for safety in today's traffic, where other drivers will probably mistake hand signals as mere gestures.

It's quite easy to convert the parking lights and taillights to function as signal lights, as they do on more modern trucks, especially if you specify the signal light option (provided by some harness makers for mid-fifties and earlier trucks without standard signal lights) when you order the wiring harness. It will still be necessary to convert the parking lights and taillights to double-filament bulbs and to add a signal light switch, if your truck doesn't already have one, but having the signal light wiring in the harness avoids botching up the new wiring with what is bound to look like Christmas tree strands to the signal lights. If you decide to use the parking lights and taillights as signal lights, you may need to add a passenger's side taillight. Through the mid-fifties, Ford pickups came standard with only one taillight, and it was on the driver's side.

Besides signal lights, you should look at other options that may be available with the harness as well. Some suppliers offer a choice of fabric- or plastic-covered wiring. If your truck had fabric-covered wiring originally, and your concern is originality, then fabric wrapping is the choice. But if you are restoring or rebuilding your truck to drive, plastic-coated wire is a better choice because it will be more durable. If your truck has a heater that ties into the electrical system, you will also need a heater harness. Accessories such as overdrive—available on fifties-vintage Ford trucks—also require a separate harness.

Installing a harness isn't just a matter of connecting the correct wires to the terminals on the gauges and other electrical components. It is also very important to be sure that the contacts are clean and that circuits, like the lights, that ground through the body sheet metal have a good ground connection. Electrical systems in most older cars and trucks used the body metal and frame for ground. This ground path is easily broken by rust or, on restored trucks, by heavy coatings of paint. To ensure a good ground, paint or surface rust needs to be scraped away from the spots where electrical components come in contact with body and frame metal. New or replated bolts should be used in attaching these electrical components.

Before you install the new harness, it's a good idea to buy new grommets. The grommets are used to plug holes where the wiring passes through the firewall and other locations in the sheet metal. Wiring grommets are available from the vintage Ford truck parts suppliers listed in the Appendix.

Harness Installation

Where you begin installing the new harness is arbitrary. You can start at either end (at the headlights or the taillights) or in the middle (at the instruments). The portions of the harness that run to the lights and under the hood can be installed by disconnecting the old wires and replacing them with new, one terminal at a time. As you remove the old harness, you can run the new wiring in its path. This process is easy and straightforward, provided the wiring has not been modified greatly. With a frame-

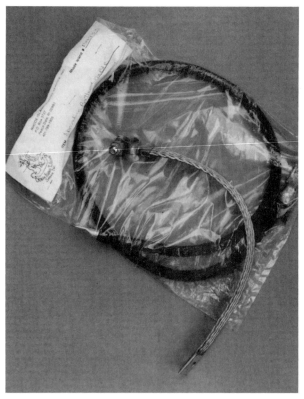

In replacing the wiring, don't overlook the battery cables. Six-volt electrical systems require thicker cables than those used with 12 volts. Replacement cables should be purchased from a reputable wiring harness supplier such as Rhode Island Wiring Service. Above all, don't replace the battery cables with versions from a discount mart, which will probably be marginal on a 12-volt system and will not carry the current needed by a 6-volt system.

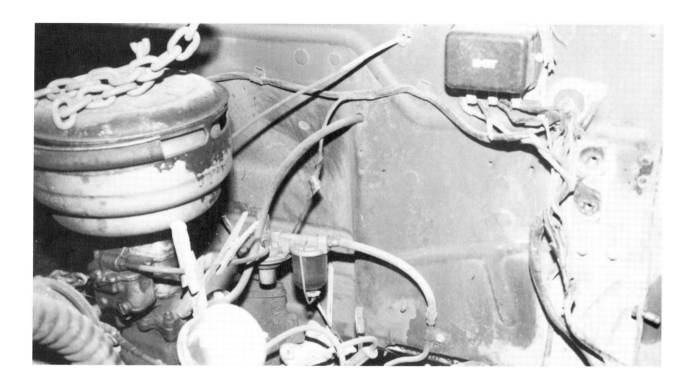

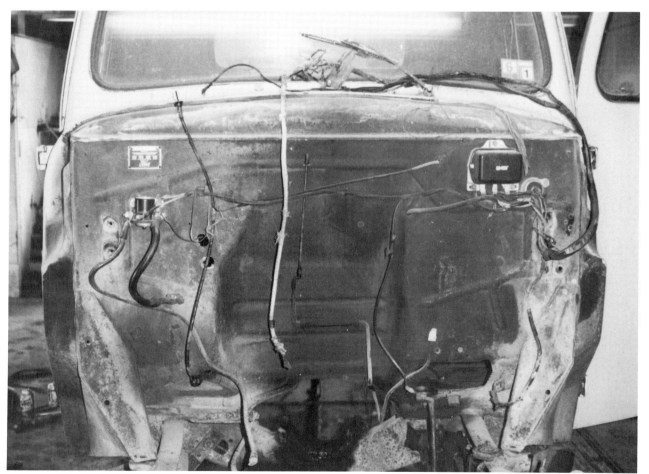

The old wiring can be removed as your truck is disassembled. If you are doing a frame-up restoration, take plenty of photos to show where the wiring was routed.

up restoration, the old wiring will have been removed long ago. In this case, you will have to rely on a wiring diagram (found in a service manual for your year and model truck) and the harness manufacturer's instructions.

Dashboard and Cab Connections

The most difficult connections are under the dash. This is because so many wires crowd in a small space and also because you almost have to be a contortionist to get yourself in a position where you can see the connections and remove and attach the wires. You'll find that removing the seat cushions and sliding the seat frame all the way back make the job of lying on your back and working above your head slightly more tolerable. In addition, banging your head on the clutch or brake pedal is something you will not want to do more than once or twice. On Ford trucks that place the gas tank under the seat, it's also necessary to remove the seat to replace the wiring to the fuel level sending unit.

Even though hooking up the dash wiring may seem to be a very complex job, if you carefully follow the instructions with the harness, plus the wiring diagram in the Ford shop manual, you'll find that this toughest part of installing the new wiring harness is actually as simple as doing a paint-by-number set. The main problems, as mentioned, are the cramped work space and the contortionist position needed to reach almost everything under the dash.

While you're hooking up wiring inside the cab, you'll also want to run the wire to the dome light (located at the back of the cab in 1953 and later Ford pickups) and the wire to the fuel sending unit on trucks that carried the fuel tank inside the cab.

Engine Compartment Connections

With the dash and cab wiring out of the way, the next portion of the harness you'll be connecting is inside the engine compartment. Compared with those under the dash, the connections here will be very easy. If the old harness is still in place, it's simply a matter of disconnecting one wire and connecting its replacement on the new harness.

If you plan on painting or detailing under the hood, this should be done before installing the new wiring. You will also need a set of grommets to plug the holes where the harness passes through sheet metal. This gives the new wiring a finished look and prevents chafing, which can eventually cut through the harness wrapping and create a short circuit or burn out the electrical system.

Headlight and Taillight Connections

You'll notice that the main wiring harness does not run all the way to the lights. For the headlights

An automotive circuit makes a complete loop from and back to the battery. In most cases, the return path is through the chassis or body metal, as shown in this diagram.

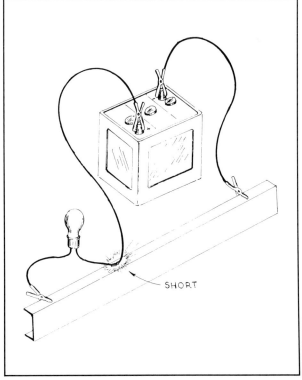

If the wiring has a short or the return path has a break, the circuit will not power its electrical components.

and parking lights, a separate harness is used. This is also the case with the taillight or taillights. Both of these light harnesses connect to the main harness with plug connectors.

If you've been working around the front of the truck, connecting the headlight and parking light harness section is the next logical step. To replace the headlamp wiring, you'll need to remove the retaining ring and headlight.

With the light disassembled, it's a good idea to inspect the condition of the buckets. Since the inside of the fenders was open on these older trucks, mud and road salt often collected on top of the headlight buckets, and you may find that the bucket assembly needs to be sandblasted and repainted. If your truck has a lot of rust, the headlight buckets may also be rusted and need to be repaired and replaced. When replacing the headlight buckets in the fenders, it's a good idea to install new rubber gaskets between the buckets and the fenders. Most vintage Ford truck parts suppliers also carry replacement rubber boots to cover the hole where the wiring enters the back of the headlight buckets. Usually the old boots are deteriorated and should be replaced.

After connecting the wiring to the headlights on both sides, you can turn your attention to the parking lights. Here, too, you will need to remove the lenses. If you decide to convert the parking lights so that they function as both parking lights and turn signals, you will also be fitting these lights with double- rather than single-filament bulbs. Dennis Carpenter sells the needed double-contact socket and wire assembly to convert parking lights on 1948–56 Ford trucks to work as turn signals.

With the new sockets in place, and with double-filament parking light–turn signal bulbs installed, you can replace the lens housings. If the lens gasket is deteriorated, it should be replaced as well. The parking light–turn signal conversion is recommended for several reasons, the main ones being the preservation of the truck's original look and the importance of signal lights in today's driving conditions.

Solenoid and Starter Connections

Now it's back inside the engine compartment, where you'll be hooking up the wiring to the solenoid and starter—if you haven't already done this. On Ford trucks with the flathead six-cylinder engine, the starter and generator are on the left-hand, or driver's, side. With a flathead V-8, they're on the right.

Once again, if the old harness has been left in place, replacing this wiring is a simple matter of disconnecting one wire and connecting another. If the harness has been removed, you will need to follow the instructions that came with the harness, with perhaps reference to a wiring diagram.

Dimmer and Brake Light Connections

Next it's down and under. The first stop is the dimmer switch. This is a simple hookup, but watch out if the dimmer switch has been replaced with an aftermarket switch. If it has, the wiring hookups may not match the harness instructions or the wiring diagram. If the switch isn't a Ford part, replace it with a quality reproduction switch from a vintage Ford truck parts supplier.

The brake light switch is also a simple connection. If this switch is corroded, or if your truck's brake lights haven't been working, it's a good idea to replace the switch—and now is the time. Older Ford trucks used a hydraulically operated

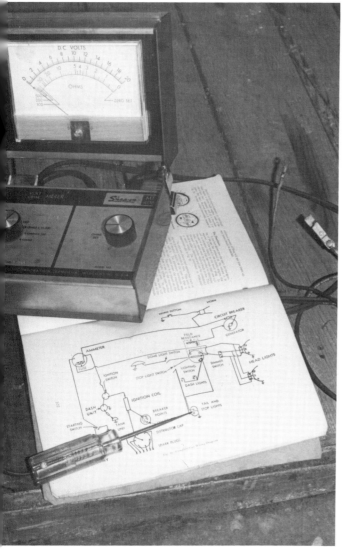

To troubleshoot a nonoperating electrical component, you will need a wiring diagram for your truck and a test device. A VOM, shown here, works best because it indicates not only that the circuit is passing current (continuity), but also the amount of current flowing to the component.

brake light switch that screws into the rear of the master cylinder. Replacing this switch is a simple matter of screwing out the old unit and screwing in the new. You won't lose any brake fluid and won't have to bleed the brake system in the process. These switches can be obtained through your local NAPA auto parts store.

Rear Light Connections

For running the harness to the rear lights, many harness makers provide some extra wire so that you can route the harness as best suits your needs. Typically this harness runs along the driver's-side frame rail and attaches to the frame with clips. New clips may be included with the harness. If not, and if the clips from your truck are broken or missing, you may want to order replacements from your Ford parts supplier.

Hooking up the rear lights is not overly complicated. Two taillights were an option, not standard, on Ford pickups into the mid-fifties, but over the years, many of the early trucks have acquired the second light. If the taillights on your truck have been replaced by aftermarket units and

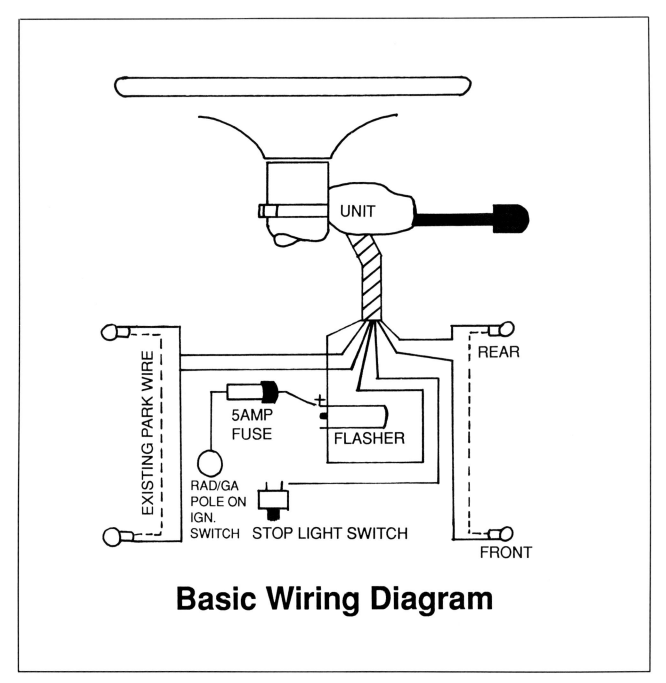

Basic Wiring Diagram

you want to return to the original-style light, Dennis Carpenter and other vintage Ford parts suppliers have the correct replacement lights, as well as mounting brackets.

Back here, it's not necessary to change the sockets and bulbs, because the taillights are already wired for stoplights and running lights and contain double-filament bulbs. The signal lights will use the brighter stoplight filament.

Turn Signal Switch Connections

The wiring harness is now hooked up, but a couple of details may still be left to finish. If you have made the signal light conversion, one step that remains is installing the turn signal switch. Correct vintage-looking switches that attach to your truck's steering column and can be tied into the wiring harness—if you installed a harness that included the signal light hookups—are available from your local NAPA parts store or the vintage Ford parts suppliers. If you installed a harness that did not include the signal light wiring, but made the necessary modifications at the parking lights, you will need to run the extra wire to hook up the parking lights and taillights to function as signals. (See the "Turn Signal Installation" sidebar for instructions on doing a signal

---

Turn Signal Installation
by Jim Simpson. Reprinted with permission.

Installing turn signals in a forties or early-fifties Ford is not difficult—I did it, and I never tried any serious electrical wiring on a car or truck before.

I decided to install turn signals on my 1950 Ford back when the electrical system was still 6-volt; it is now 12-volt. The decision was one I sort of eased into—I had thought about it many times, but never quite got motivated to do it. The safety aspect was my primary goal. After a few near misses by other drivers, I decided it was time to get it done so other drivers would know which way I was turning. (Amazing how many drivers these days don't know how to read arm signals out the window.)

Installing turn signals is a relatively simple task and takes only 2 to 4 hours. The total cost for my signals was under $25—a cheap enough investment for the peace of mind it produces. Following is a list of materials I used; you might find it helpful. After that is a step-by-step run-down of how I wired the system.

Materials
- Basic turn unit. I purchased an Everlasting-brand unit, part number HL-102, at a swap meet for $17. The average cost in catalogs is $26 to $29. This is a chrome unit that attaches to the steering column with a clamp. It has color-coded wires and an easy-to-follow diagram. A green light in the end of the handle tip flashes when the signals are engaged. This is a manual unit; it does not cancel automatically.
- Three-prong flasher and holder. This is simply a clamp that holds the signal flasher up under the dash.
- Dual-element bulbs for the front parking light units.
- Dual-element (or two-wire) sockets to replace the front single-wire sockets.
- In-line fuse holder with 5amp fuse.
- Collection of electrical connectors and soldering gun.
- Electrical tape.

Installation

1. Attach the basic turn signal unit to the steering column near the steering wheel.

2. Run new wire from the proper wire at the unit to the driver's-side parking light unit and wire it and the parking light wire into the new two-wire socket, replacing the single-element bulb with the dual-element bulb.

3. Run new wire from the proper wire at the unit to the passenger's-side parking light unit and wire it and the parking light wire into the new two-wire socket, replacing the single-element bulb with the dual-element bulb.

4. Run new wire from the proper wire at the unit to the driver's-side taillight unit.

5. Disconnect the original crossover wire that leads from the driver's-side taillight unit to the passenger's-side unit. Splice a new wire to the crossover wire and attach this wire to the appropriate passenger's-side wire at the unit. Attach the driver's-side wire to the driver's-side unit where the crossover wire was disconnected.

6. With both front and both rear wires connected (you can use butt connectors or, better yet, solder the connections and cover them with shrink tubing for a neat job), connect the appropriate wire from the unit to the brake switch at the master cylinder. (This will leave one original wire to the brake switch unattached. Tape it off at the switch or cut it at the wiring harness under the hood and tape it.)

7. Attach 5amp fuse holder wire to the radio–general accessory (RAD/GA) pole on the ignition switch (or other live pole). Attach the other end of the fuse holder wire to the positive tab on the flasher. Connect the other flasher wires following the directions in your particular kit, or use the accompanying diagram.

8. Turn on the ignition key and test the lights. You should have blinking lights both front and back.

Now try the opposite side. If all is well, turn the key to the on position and repeat the test. You should have turn signals front and rear.

light conversion when the signal light wires are not part of the harness.)

If your harness included the signal light circuit, you'll find that hooking up the signal lights is easier than it looks. The wiring instructions that came with the switch are color coded, and the instructions with the harness are typically numbered. It is just a matter of matching these two instructions. For example, the harness may list wire number one as the left-turn front signal lamp, and the instructions with the switch may show the left-turn front signal lamp wire as black. This means you will connect the black wire from the switch to the number one wire from the harness.

Horn Connections

The last detail is hooking up the horn. This is a single-wire connection, but the horn button must be removed. Usually the horn button comes loose by pushing down evenly on the button and turning it counterclockwise. The new horn wire is pulled up through the steering column by tying it to the old wire.

Even though this may be your first attempt at anything electrical, with the wiring harness instructions and a shop manual, the project will be much easier than it might appear when you first lay out the new harness. From start to finish, if you have all needed parts in hand, you can expect the project to take between 16 and 20 hours—less if you have had previous experience.

In addition to making your truck safer to drive, installing a new wiring harness gives the additional bonus of the satisfaction found in a job well-done. Replacing the bare and spliced wiring in your truck is an investment you will never regret.

Electrical Problem Troubleshooting

It is fairly likely that even though you have just installed a new wiring harness, some electrical component won't work. Maybe you confused a couple of the wiring connections, or maybe you didn't clean the terminals thoroughly enough. Maybe the component is not making a good ground, or maybe a problem exists in the electrical component itself. To fix these electrical problems, you'll need to know some basic principles of

A 6-volt alternator from Fifth Avenue Antique Auto Parts can keep your truck's battery fully charged, even with occasional driving.

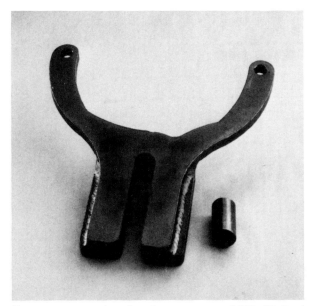

If you are mounting the alternator on a 1932–53 Ford flathead V-8, you will need this special Y-bracket. Randy Rundle

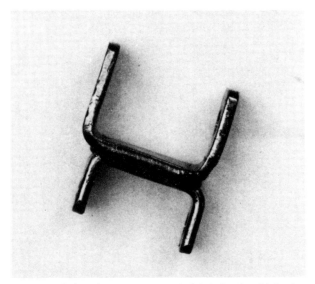

I mounted the alternator on a Model A Ford, which also required a special bracket. Fabricating a mounting bracket is no big job; just bend some strap iron, drill holes for the mounting bolts, and you're in business.

Reenergizing Your Old Truck's Electrical System

It's no fun to slide behind the steering wheel of your collector truck, push the starter button, and have nothing happen. Darn, dead battery! If your garage is on a hill, maybe you can push the vehicle out the door and let it kick itself to life as it rolls down the grade. Otherwise you'll have to wait for the battery to charge, or haul out the jumper cables, bring over another battery, and do a jump-start. It sure is a lot easier just to have the starter kick the engine to life.

The main reason for the low- or dead-battery plague is that most of us don't run our collector trucks often and long enough for the generator to recharge the battery fully. The blame really lies not with the pampered use we give our "pets," but with the inefficiency of the generator that is the heart of our older truck's electrical system.

Generators do a good job of keeping a battery charged during frequent stints of distance highway driving. They do a poor job in short, low-speed runs—the pattern of much of our hobby driving. So what's to be done? How about a once-a-week 100-mile cruise? Not practical? Well the other option is to hook up the battery charger every ten days or so. Sounds tedious, and we'll probably forget. Wait, there's another solution: How about replacing that lazy old generator with a spunky, high-output alternator. Since an alternator produces as much electrical energy at low speeds as a generator does at high speeds, using an alternator to energize your Ford truck's electrical system will keep the battery at full charge just from occasional around-town runs.

Why am I hearing objections? So an alternator's not authentic to your older vehicle. Keep the generator and bolt it back in place when you're going to a show. Oh, that's right, some of our older trucks are running 6-volt electrical systems, and alternators didn't show up until the sixties when all cars and trucks were running 12-volts. True, but you can buy a 6-volt alternator from Fifth Avenue Antique Auto Parts, and this little dynamo puts out a 50amp charge. Drive across town to visit Uncle Bob, and that high-output alternator will have stuffed enough electrical zip-zowie into the battery to spin the starter like a windmill in a hurricane. Willing at least to consider the alternator route? Let's look at how simple the swap is.

First you'll need an alternator. For a 12-volt system, you can use an alternator from a modern car or truck, but you'll need to buy a DA plug so that you don't have to change the ignition switch. Otherwise the alternator cannot be deenergized; it won't shut off and will keep the engine running when you turn off the ignition switch. If your vintage vehicle has a 6-volt electrical system, you'll need to buy the 6-volt alternator.

Wiring the alternator into the system is simple. Chances are, the biggest challenge will be engineering a new mounting bracket. If your Ford truck is a 1932–53 with the flathead V-8, you'll want to buy a special mounting bracket—also available from Fifth Avenue Antique Auto Parts; otherwise you will adapt the generator bracket or make a new one. The installation shown here is on a Model A Ford, and since the generator has its own integral bracket, a new bracket had to be fabricated. Making a bracket is no big job; just bend some strap iron (available from a good, old-fashioned hardware store), drill holes for the mounting bolts, and you're in business. Be sure to make the bracket yoke slightly wider than the mounting point on the alternator so that you have some room to adjust the pulley alignment with the fan belt. If your vintage Ford truck has the old-style wide fan belt, you will also need a wide pulley. No problem; Fifth Avenue has these, too.

Remove the generator by loosening the bracket that adjusts the belt tension, then remove the mounting bolt. Install the new mounting bracket, or put a spacer in the old bracket if that will suffice, and mount the alternator. Make sure the pulley aligns with the fan belt, pull back on the alternator until the belt is tight, and reattach the belt tension bracket. Now all that's left is the wiring.

Piece of cake; just disconnect the battery (Batt) wire from the voltage regulator and attach it to the back of the alternator. You'll need to run another wire from the alternator's white wire to the hot side of the ignition switch or Batt side of the coil.

Now remove the old voltage regulator, and the job is done—unless the electrical system is positive ground. The alternator is negative ground, so you'll need to reverse the polarity. To do so, turn the battery around in its box so that you can attach the positive post to the starter cable and the negative post to ground; you'll get better starter response if you connect the ground cable to a starter bolt and not the frame. Now reverse the wires on the coil and the amp gauge.

That does it. The electrical system is now negative ground, and its energy source is a high-output alternator that'll keep the battery fully charged, even on short hops. Just make sure the battery cables are clamped tightly to the battery posts before you head down the road. Should a clamp work lose, you'll probably burn out the diodes in the alternator—requiring a big-money repair. If you make a practice of loosening a battery cable when your vintage vehicle is sitting for a period of time, buy a battery switch and install it instead. That way, you won't risk a cable coming loose on the road.

Converting to an alternator-energized electrical system won't cost you a lot of money—maybe $150 maximum—and you'll save yourself that dead-battery disappointment, plus all the hassle of keeping the battery on-line for those times a friend stops by and you say, "Want to go for a ride in my old truck?"

The actual installation on your truck is a piece of cake. The hookup uses existing wires.

electrical troubleshooting and to have an understanding of how electrical circuits operate.

To work through the troubleshooting approach and see how to trace down a problem in an electrical circuit, let's suppose that one of the parking lights doesn't work. That's what happened when I installed a new wiring harness in my truck. Both headlights worked, but only one parking lamp would light up.

Electrical problems have three possible causes: power is not getting to the component (in this case the light) or the circuit is not being completed owing to a poor ground or the component itself has failed. Using the example of a nonfunctioning parking light, we'll see how to figure out the cause of our electrical problem and learn how to fix it.

First we'll turn on the parking lights at the switch on the dash. Since one parking light comes on, we know the switch is working and current is flowing in the circuit. Let's assume the parking light that's working is on the driver's side. Since the passenger's-side parking light attaches to the wiring harness after the driver's-side light, we can only be sure that the circuit is good up to the driver's-side parking light. So we're back to our three potential causes for the passenger's-side parking light's not working: current may not be getting to the light or the circuit has a poor ground or the light is burned out.

When electrical troubleshooting, it's important to know that every problem has a logical cause. Even though it can't be seen, electrical current is not some magical energy that either works or doesn't work, as it pleases. To find the cause of the dead parking light, we begin by removing the retainer rim and the glass or plastic lens. This is done so that we can check to see if current is flowing to the light socket. If we get a current reading at the socket, we know two things: the wiring is good to this point, and the circuit has a good ground. If we eliminated these two causes, the problem would have to be a burned-out light bulb.

Two devices can be used to check current flow. One is a simple test light that you can buy for a few dollars in the automotive section of a discount mart or an automotive parts store, or make yourself from an automotive light socket and two short sections of wire. The other test device is a VOM. A VOM is preferred to the test light because it also shows the amount of current the circuit is carrying. This meter is available from discount marts and electronics

To prevent any possibility of a short or electrical fire during storage, the electrical circuit to the battery should be disconnected. The best way to do this is with a quick-disconnect switch. This switch also serves as a theft deterrent.

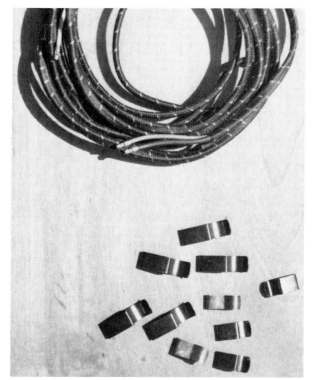

Along with the wiring harness, order a supply of clips. These slip over the frame and hold the harness in place.

Electrical Shutoff

Replacing frayed and decayed wiring certainly gives great peace of mind, but even so, it's wise to disconnect the truck's battery from the electrical system during storage—even if the vehicle will only sit for a few days. If your truck has an original electrical clock, it is imperative to disconnect the battery because the clock will run down the battery in a relatively short period of time and burn itself out in the process. Even on trucks without a clock, it's wise to disconnect the battery because of the possibility that an unnoticed trickle discharge may occur somewhere in the system and not only may discharge the battery, but could start a fire.

Rather than loosen and remove a battery terminal each time you park the truck, the easier method is to install a quick-disconnect switch on the battery terminal. If you plan to enter your truck in shows, this switch can easily be removed at show time.

With the switch in place, it's just a matter of getting in the habit of turning it when you park your truck. You will also find that the switch provides a small measure of antitheft protection, since would-be thieves may not notice or want to take the time to bypass the switch.

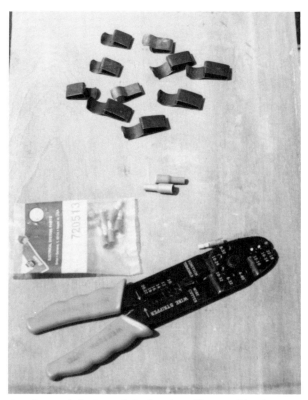

Another tool that's useful when installing a wiring harness is the insulation stripper shown here. Although the harness will come complete with all connectors, it may be necessary to strip off wiring insulation to install or replace connectors on some of your truck's electrical accessories.

stores for $25 or less.

Whether you are using a test light or a VOM, the hookups will be the same. Place the hot lead (the red wire from the commercial test light or meter) against the contact point in the bottom of the light socket, and hold the other lead (which will be black) against a ground. At the parking light, a ground connection could be made with the light assembly or some unpainted spot on the fender or bumper.

Getting a good ground is important. If the test light or meter does not make contact with ground, then it won't show current flow, even if the current is actually passing to the light socket. With a good ground, if the test light illuminates or the meter hand swings into the appropriate current range on the dial—6 volts for a 6-volt electrical system or 12 volts for a 12-volt system—you know the problem is that either the light bulb is burned out or the light itself isn't making a good ground.

You can check for a poor ground in the parking light circuit by connecting the black lead from the test light or meter to the outer casing of the bulb socket. If the light goes on or the meter needle moves, current is reaching the socket and the problem is with the bulb. If the light or meter doesn't show current, the socket is not making a good ground.

You can correct for a poor ground in two ways. One is to remove the light housing and clean the metal-to-metal contacts between the parking light housing and the fender so that some bare metal is showing, then reattach the light housing. If the

Conversion to 12 Volts

One popular conversion to older trucks is upgrading the electrical system to 12 volts. This upgrade is done for a number of reasons: 12 volts give easier starting, the modification provides more power for modern accessories like air conditioning, replacement electrical parts are more readily available for the higher voltage, and the conversion is simple. To make the change, you will use the 6-volt wiring and starter. You will need to replace the battery, generator, regulator, and light bulbs; you will use a 12-volt flasher if your 6-volt truck has turn signals; and you will place a voltage resistor in-line with the dash gauges and radio.

The 12-volt battery will probably require a different battery box. Since 12-volt systems are negative ground, you will connect the battery cables so that the negative battery post goes to ground. Most 6-volt systems are positive ground.

Ford converted to 12 volts in 1956. If you can locate a 1956–63 or later Ford generator, you can probably use it, though you may need to replace the end frame on the 12-volt generator with that from your 6-volt. Another alternative is to replace the field coils in your truck's 6-volt generator with those from a 1956–63 Ford unit. The field coils can be interchanged in Ford generators back to the thirties. A third approach is to have the generator rewound for 12 volts.

The 12-volt regulator and coil can also come from a 1956 or later Ford. A voltage resistor is needed to drop the voltage for the dash gauges and radio. If your truck has a heater, you will either need to put a resistor in-line with the heater motor or replace the motor with one from a 12-volt Ford car or truck. Since the polarity of the electrical system has been changed, you will need to reverse the wires on the ammeter.

All light bulbs must be replaced with their 12-volt counterparts. It's unlikely that the sockets will require modification. The starter should work fine on 12 volts without any modification, and will crank the engine over much faster than with 6 volts, for easier starting.

If you use stock-appearing parts, the only giveaway will be the extra three cell caps on the battery.

Wiring harnesses typically come in several segments. Replacing the taillight harness is easier with the truck's box removed. This harness segment also holds the wires to the fuel gauge sending unit.

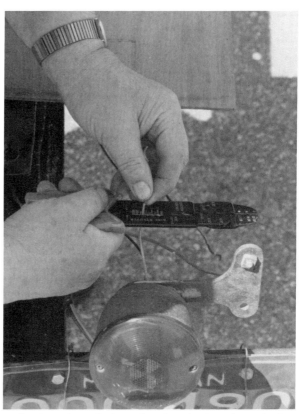

To hook up the taillight, a short length of insulation had to be stripped off the wires so that plug-in connectors could be crimped onto the wire ends. The stripping tool makes this an easy job.

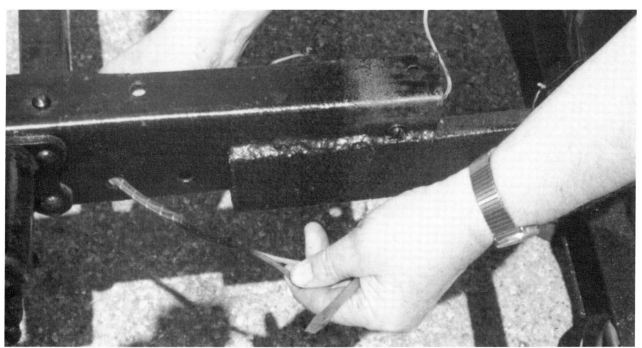

The taillight leads need to go outside the frame through one of the holes in the rear frame extensions. On some harnesses, a length of flexible metal conduit is supplied to prevent the harness from chafing against the frame and wearing through the insulation.

With the plug-in ends installed on the taillight wires, hooking up the taillight was just a matter of fitting the plugs at the ends of the harness leads, as well as the new plugs in the taillight wires, into short female connectors. These connectors were not supplied with the new harness and had to be salvaged from the truck's original wiring.

socket contacts are dirty or corroded, you should also clean the socket, using emery cloth or fine-grit automotive sandpaper. Now test the circuit again, this time grounding the test light or meter to the socket casing. If the test device works, the problem is corrected.

Sometimes it is difficult to create a good ground through sheet metal. When this happens, you will use the second method for ensuring a good ground. This requires stripping the ends from a length of wire and attaching one end to the parking light casing, in this case, and the other end to a ground point on the chassis. You won't want to use this method when doing a show restoration because the ground lead will be nonoriginal and will cause your truck to lose points. But on a truck that is being overhauled so that it can be enjoyed and driven, running a ground wire is an easy way to ensure a properly functioning circuit.

If the problem isn't one with the ground or the component (like a burned-out light), then the circuit has a break. In this case, you need to determine where the break occurs. This is done by tracing back along the circuit, following the harness and referring to a wiring diagram, until you find a plug connection, switch, or some other accessory powered by the same circuit. At this point, you will

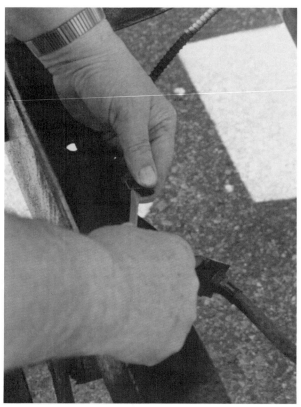

With the taillight connected, we could attach the harness to the frame. This was done by fitting the clips over the harness, then pushing them onto the frame rail.

again test for continuity—to see if power is flowing to that point.

With the parking lights, tracing the circuit back will take you to the plug connections between the main harness and the short harness that goes to the lights. You will disconnect the light harness and check for current at the sockets from the main harness. You make this check by placing the red lead of the test light or VOM on the hot wire and the black lead on ground; to find which wire is hot, you will need to trace the circuit to see what color wire goes to the parking lights from the headlight switch. If current is flowing to this point in the main harness, then the problem is probably a corroded plug connection or poorly made contact between the harnesses.

Although the troubleshooting process will vary with different electrical components (in some cases, the problem may be at the switch or switch connections), following a logical and methodical step-by-step approach, you will be able to trace the source of any electrical problems and make the necessary repairs until all of your truck's electrical components are working properly. Keep in mind that electrical accessories like the horn, clock, heater motor, fuel sending unit, or radio may well have quit working long ago and will thus have to be rebuilt or replaced.

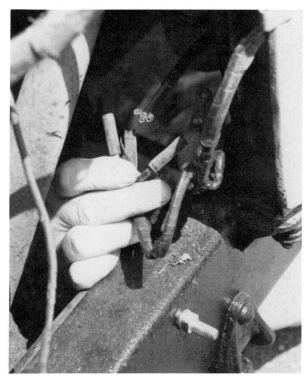

At the front, the taillight harness plugs into leads from the main harness. Here, too, the connectors had to be salvaged from the old wiring.

Chapter 13

Redoing the Interior

Although truck interiors are designed for utility, ruggedness, and durability, a vintage Ford truck would have to be extremely well-cared-for not to show signs of wear and perhaps some tears on the seat, a sagging headliner, and other evidence that the interior needs to be upgraded or replaced. Fortunately redoing a truck interior is a much simpler and less expensive process than replacing the upholstery in a car.

Until the mid-fifties, Ford truck seat coverings consisted of plain, durable Naugahyde or vinyl; headliners were simple cardboard strips; door panels, when not steel, were vinyl-covered cardboard; and the cowl kick panels were likewise vinyl-covered cardboard, or plain steel in more modern trucks. Through the mid-forties, even the color of the interior was durable. On Bonus Built and earlier Ford trucks, seats and cardboard interior trim were a sort of moose brown color that withstood soiling because dirt and grease blended right in. But that's not to say that a worn, soiled interior looks good in a refinished truck. One of the most dramatic improvements you can make is to re-cover the seats, refinish the interior, and install new headliner, door, and cowl panels.

Interior Replacement Kits and Supplies

Thanks to seat covering, headliner, and door panel kits, which are available for most 1928–66 Ford light trucks, redoing your truck's upholstery can be a highly satisfying do-it-yourself undertaking. If you'd rather have an upholstery shop do the job, it still pays to work from a kit for two reasons. First, with a quality interior kit from one of the reputable

The most complete line of upholstery kits for Ford cars and trucks is made by LeBaron Bonney Company in Amesbury, Massachusetts. Lee Atherton, the company's founder, stands with samples of the upholstery materials.

The first step in replacing seat upholstery is to remove the cushions and strip them of whatever remains of the old covering, padding, and spring wrapping. If the spring sets are badly rusted, you have two alternatives: find a better set or rebuild the set you have.

Ford light-truck parts suppliers listed in the Appendix, you'll come as close to matching the original color and materials as it is possible to get. Second, besides keeping your truck as close to original as possible, a kit enables a professional upholsterer to replace the seat coverings and install the interior panels in a fraction of the time that would be required to make these items from scratch. If your upholsterer has had a bad experience with an automotive interior kit, he or she will be understandably leery of the kit concept. You can assure your upholsterer that shoddy materials and artisanship will not be found with a Ford truck interior kit purchased from a reputable supplier.

A complete pickup interior consists of several items that are typically sold separately. These include the seat coverings, new coils with which to rebuild the spring sets, headliner, door panels, kick panels, sun visors, armrests, and related hardware.

Tools

Like other phases of your truck's restoration, interior work calls for a few special tools. These include heavy-duty scissors; hog ring pliers, for attaching the seat upholstery; a utility or Xacto knife; a tool for removing the window cranks; and a heavy-duty wire cutter.

Preparation

Seats are usually removed from the truck during body repair and painting, so they can be re-covered anytime. Installing the headliner and door panels will wait until after exterior and interior painting. If your truck still has the original cardboard headliner, door, and cowl panels, take close-up photos or make sketches of exactly where these pieces fit and how they are in place. If these trim pieces have been removed from your truck, it's a good idea to take photos of correctly installed interior trim in restored trucks. These photos will be helpful when it comes time to put the new interior trim pieces in place. It's a good idea as well to keep the original pieces to use in checking the authenticity and fit of the new panels. The original headliner can also serve as a guide when bending the replacement to fit the roof contours.

Seat Cushion Re-covering

Several preparation steps need to be done before actually re-covering the seats. First the cushions will be removed from the truck and stripped of whatever remains of the old covering, padding, and spring wrapping. To take off the old covering, use cutter pliers to cut the hog rings that hold the cover to the

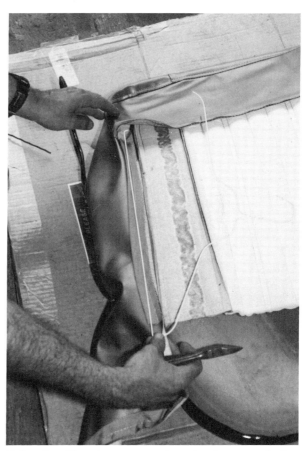

Before installing the seat covering, check to see whether cording is sewn into the seams at the edge of the cover. If not, wires salvaged from the old cover—or replacements if the original wires are missing—need to be slipped into the seam at the bottom of the cover flaps. Replacements can be cut from a coat hanger.

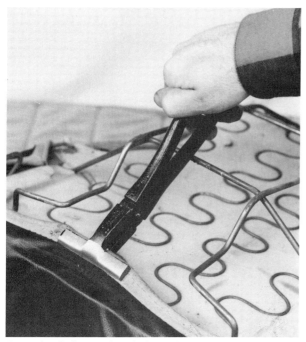

Special upholstery pliers help pull the fabric tight over the spring frame. The Eastwood Company

167

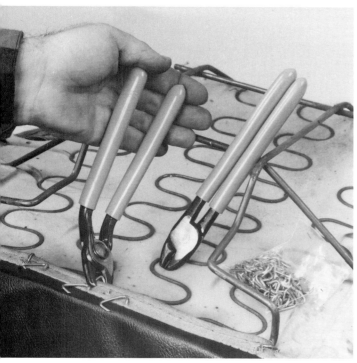

The seat cover attaches to the frame with hog rings. Although these rings can be installed with regular pliers, a special upholsterer's tool called a hog ring pliers, which has grooves in the jaws to hold the hog rings, makes this job much easier. The Eastwood Company

seat frame. If the seat cushion is wrapped with burlap, you may also find hog rings holding this wrapping in place.

Don't discard the old seat coverings, padding, or burlap. The covering may have wires sewn into seams along the edges; these are called listing wires, and you'll need to reuse them. If the seats have foam padding, you may need to reuse the form or, if it is badly deteriorated, use the original as a pattern to cut a new foam cushion. You may also decide to reuse the burlap wrapping.

With the old covering removed, you can inspect the spring sets for soundness. On a truck that's seen a lot of use and abuse, it's not uncommon to find breaks in the seat frame as well as broken or badly sagging springs, and a mild to severe coating of surface rust. If the spring sets are badly deteriorated, you have two alternatives: either find a better set or rebuild the set you have. On Bonus Built and newer-series Ford trucks, it should not be difficult to locate a set of sound, re-coverable seats at a scrap yard or through want ads placed in a club newsletter. For late-thirties to mid-forties Ford trucks, finding replacement seat springs may not be as easy. New seat springs are available for Model As and Model Ts. Spring sets can be rebuilt by welding cracks or breaks in the frame and replacing the broken or sagging coils. New seat spring coils are available from some of the parts suppliers listed in the

Attach the seat cover at the front first. To do so, pull the cover down until the end flap is even with the rod or channel at the bottom of the seat frame. Next insert a hog ring into the hog ring pliers, press the hog ring over the fabric and seat frame, and squeeze the pliers.

Most of the interior surfaces on an older truck are painted. Quite a bit of preparation work is entailed in repainting the interior, and this job needs to be done before the seats are replaced or the headliner, door, and kick panels are installed.

When you are refinishing the cab's interior, pay careful attention to detail. The result will be a truck you can display with pride. John Csaftis

If you need to change the coloring of the fiberboard headliner, door, or kick panels, the heavy cardboard material can be spray painted.

Appendix. If the spring frames are rusted, you should clean the metal by sandblasting or with a wire brush, and spray on a protective coating of rust-resistant paint. If you haven't already done so, this is the time to order the seat covering kit.

It doesn't make any difference which you recover first: the seatback or the seat bottom. Recovering the spring set can be done by one person, but an extra pair of hands makes the job much easier, especially when it comes to pulling on the new covering. The first step is to replace the burlap wrapping. Sometimes new burlap is supplied with the kit. If it isn't, the original burlap can be reused if it is still in good condition, or a new piece of burlap can be purchased from an upholstery shop. The purpose of the burlap wrapping is to keep the padding from settling into the coils. The burlap attaches to the seat and backrest frames with hog rings that can be purchased either at an upholstery shop or at a farm supply store. Hog rings from the farm supply are made of a thicker metal—because they're intended for use with animals—but work just as well as the thinner upholstery hog rings.

Next, on early-model pickups, you will place cotton padding over the top and around the sides of the springs; later-model trucks use foam padding that sits on top of the spring set. In most cases, kits provide the right amount of padding, so don't cut off any apparent excess until you are sure everything is installed right.

On older models where the seats are padded with cotton, the kit may also contain a layer of foam rubber. If this is the case, the foam is placed on top of the cotton padding. If the foam layer is not included, you will probably want to purchase a sheet of 1in-thick foam rubber from an upholstery shop. Otherwise, as the truck is driven, the cotton will compress, lowering the cushion height and giving the seat a less comfortable feel. Foam rubber is resilient, so it will preserve the seat's cushiony feel and maintain the right cushion height for many years.

The padding and foam are held in place by the cover, which will be fitted over the cushions next. Before installing the cover, check to see whether cording is sewn into the seams at its edges. If not,

Door and cowl panels are held in place by screws and clips. On a Model A roadster pickup, the door panels slip under a lip at the top of the door and are held at the bottom with screws.

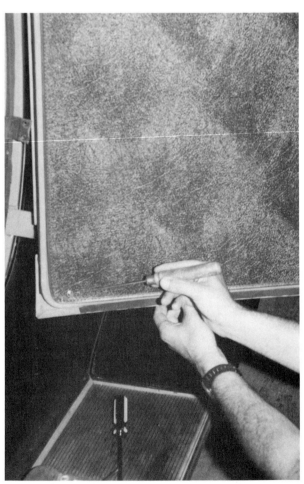

An ice pick works well to punch a hole in the door panel and to align the panel with the screw hole in the door.

wires salvaged from the old cover—or new wires if the original wires are missing or rotted—need to be slipped into the seams at the ends of the cover flaps. If the original wires are missing, replacements cut from a coat hanger make a good substitute. The original wires are usually kinked to keep them from slipping out of the seams as the cover is installed. To keep the straight replacement wires from falling out, bend a hook in one end. The hook will bind in the seam and keep the wire in place.

When installing the new covers, the cover seams must align with the edges of the cushion. To help get this alignment right, turn the cover partially inside out before placing it over the padding. While you and a partner hold the padding in place with one hand, pull the cover down over the cushion, turning it right-side-out as you do so.

When you have the cover in place, you'll probably find that the fit doesn't look entirely satisfactory. You may see puckers along the seams, the cover may be misaligned on the cushion, and wrinkles are likely to be in the fabric. To remove any puckers and to make sure the cover is correctly aligned, reach one hand under the cover and tug the cushion and padding while pulling the cover from the outside with the other hand, until the seams align perfectly with the edges of the cushion. If the padding is smooth and the cover is aligned correctly on the seat frame, any wrinkles should smooth out as you pull the cover tight. Slapping against any wrinkle areas with your hand will help smooth any minor bunching in the padding and move the cover slightly if it needs minor realigning.

If you are sure the padding is smooth and the cover seams are in alignment with the edges of the seat frame, and you still have sizable wrinkles and puckers that won't lie flat when the cover is pulled tight, chances are, you've been shipped a poorly sewn seat covering kit. At this point, it would be a good idea to contact the manufacturer and inquire about exchanging the cover for a replacement.

Sometimes upholsterers will apply steam vapor to a troublesome wrinkle area when installing a vinyl seat covering, and while the fabric is pliable from the heat, they will stretch the fabric, removing the wrinkles in the process. A steam heating wand can

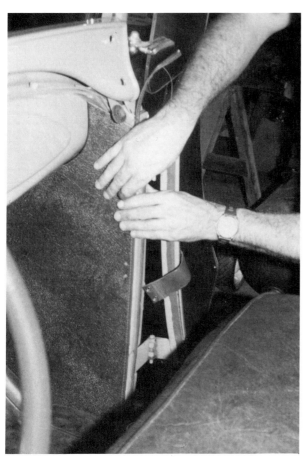

Cowl panels, also called kick panels, are installed in a similar manner. Some maneuvering may be needed for the panel to fit correctly, but upholstery pieces from a kit should not require any trimming or cutting.

On older trucks where straps are used to keep the door from swinging around and banging against the cowl, these are replaced after the kick panels are fitted.

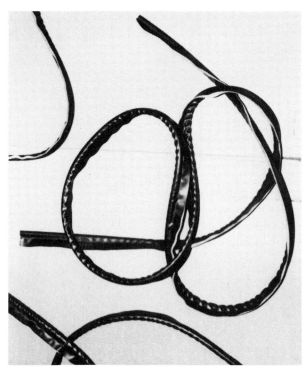

Many older trucks use lacing as weather seal along the inside of the door opening. This lacing consists of cording covered with vinyl material.

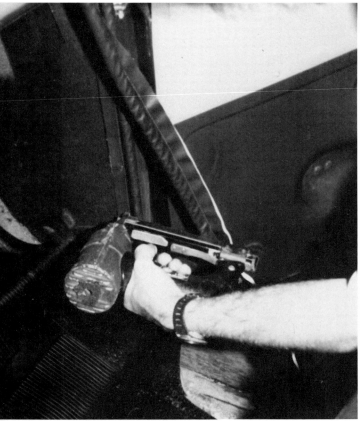

The lacing is tacked or stapled into a fiber strip that is attached to the cab metal.

be rigged by fitting a length of copper tubing into the spigot of a teakettle. This steam heating method is effective, but stretching the vinyl also weakens the material, so it isn't recommended.

Once you have the cover fitted to the seat frame and the vinyl laying smooth to your satisfaction, you're ready to attach the cover to the frame rails. Here's where the hog rings come in. The sequence of attaching the cover begins at the front of the seat frame. Pull the cover down until the end flap is even with the rod at the bottom of the seat frame, insert a hog ring into the hog ring pliers, press the hog ring over the fabric and frame rod, and squeeze the pliers. As mentioned earlier, it is important to be sure that either cording or wire is running through the seam at the bottom of the end flap. The hog rings must go around this cording or wire. If the cording or wire is missing or isn't enclosed by the hog rings, the cover will tear loose from the seat frame. Space the hog rings about 4in apart.

Next you will pull the fabric snug and clip the back flap to the rear of the seat frame with a few hog rings. If the cover has too much or not enough give, you can cut the rings at the rear of the frame and allow a little more slack, or pull the covering tighter. If you adjust the tension from the back of the seat, any holes from refitting the hog rings won't be noticed. When the covering has the desired snugness, it should be pulled slightly tighter to allow for future stretching, then clipped to the frame with hog rings spaced about 3in apart. If the cover has too much fabric when you've pulled it to the desired tightness, don't cut off the surplus, because if you do, you'll lose the seam with the support wire or cording. What you can do to take up the slack is fold the end of the cover over a couple of times. That way, you will still be clipping the hog rings over the support wire or cording and the fabric won't tear loose.

With the front and back flaps attached to the seat frame, you can secure the end flaps.

The other spring set—cushion or backrest—is re-covered following the same sequence.

With the new upholstery installed on both the seatback and the bottom, you are ready to bolt the two frame sets together—on trucks, like those in the F-100 series, where the cushions don't rest in a separate frame. This done, you can reinstall the seat in the truck and bolt it to the floor. To make sure the bolts turn easily should the seat need to be removed sometime in the future, squirt a couple of drops of automatic transmission fluid into the threads before inserting the mounting bolts.

Vinyl seat coverings clean easily and can be kept looking new with a minimum of care. A couple of words of advice may be helpful, however: Never clean vinyl with Armor All. This conditioner contains silicones that seal in the fabric so that it can't breathe. Instead wash the cover with water and a

More modern Ford trucks use more ornate interior coverings, but the upholstery panels install in essentially the same way as the older, plainer-looking revetments.

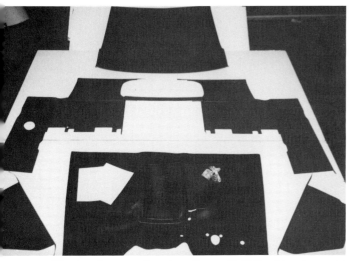

The interior coverings in a Bonus Built Ford truck consist of the headliner, side and rear trim pieces, kick panels, and floor mat. When installed, some of the interior coverings overlap one another. If you have the old interior cardboard, you can tell where the overlaps occur. If the old pieces have been discarded, you may want to take photos of the interior in a correctly restored truck. LeBaron Bonney Company

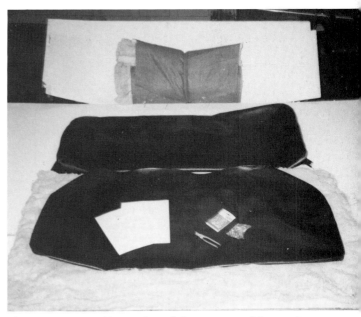

The seat covering kit consists of the cotton padding and new burlap covering for the seat springs, as well as the Naugahyde seat covering, plus installation instructions. LeBaron Bonney Company

mild soap—dish detergent works well. Seam areas will collect dirt that may not be removed with a light washing, so it's a good idea to vacuum the seams periodically.

Not only do new seat coverings make a significant improvement in your truck's appearance, but they'll probably also make your cushions feel more comfortable than they did before.

Interior Painting

The interior should be repainted before the seats are replaced or the headliner, door, and kick panels—on trucks where these items were used—are installed. The typical sequence is to paint the truck's interior before the exterior. If the truck has not been completely disassembled, extensive masking will be required to protect the windshield, steering column and wheel, and dash assembly (if it has not been removed) from overspray. If the cab has been completely stripped and removed from the frame, the recommended procedure is to paint the bottom and firewall first, then drape plastic sheeting over the frame and mount the cab. Now you can paint the interior, and hook up the mechanicals, run wiring to the dash, and so forth. The plastic sheeting will protect the frame from overspray when you paint the cab exterior and do the follow-up wet sanding process that leads to a mirror finish.

When repainting the cab interior, you will need to pay special attention to detail. On whatever year and model truck you are restoring, note the interior color and paint scheme before stripping, sandblasting, or other preparation steps. If the cab has been repainted, original paint colors should show behind the seat and in other areas that have been covered by the headliner, door panels, or cowl upholstery. Taking color photos of the interior before disassembly can be a great help not only in matching the original color scheme, but also in replacing the interior coverings.

When repainting the cab, be sure to spray a rust-resistant paint like zinc chromate or a rustproofing coating on the bare metal surfaces inside the doors. The inside surfaces of the doors can be reached

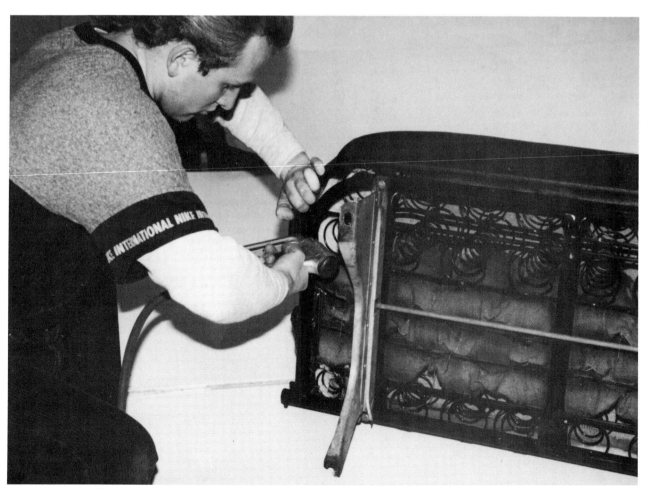

The new cover is being attached to the seat bottom. Note the covering around the seat spring coils. LeBaron Bonney Company

easily while the door coverings are removed. Left untreated, rust will continue to attack these metal surfaces.

After the cab is painted, you can install the interior coverings—just be very careful not to nick or scratch the paint.

Interior Covering Installation

Although much of the cab interior is simply painted metal, Ford gave its trucks a somewhat finished look by installing cardboard headliners; door panels; and cowl, or kick, panels. These interior coverings are available for Ford trucks through the late-fifties models in easy-to-install kit form. Whereas in car interiors, blind tacking and other tricks of the upholsterer's trade are used to conceal door panel and other trim fasteners, in trucks, trim panels are held in place in a straightforward manner with sheet metal screws.

Unlike some other manufacturers—namely, Chevrolet—Ford used a full headliner in its pickups. This makes interior roof covering somewhat tricky to install—particularly when making the bends to fit the roof contour. If you try to bend the headliner dry, you will probably crease the cardboard. The way to bend the cardboard pieces that fit against the rear corners of the cab and the large piece that forms the crown is to moisten the back side of the cardboard in the areas where the bends will occur, with a mixture of 80 percent household ammonia and 20 percent water. This solution can be placed in a Windex bottle and sprayed onto the cardboard; remember to spray the back side so that water stains will not show on the good side. The cardboard should not be soaked, just moistened. The damp cardboard will bend easily without crinkling. The ammonia may smell for a day or so until it dries.

You'll find that installing a headliner is definitely a two-person job. If you still have the old headliner, look at the pieces to see where the seams overlap. For example, on headliners where separate pieces are used above the doors, the edges of these pieces overlap both the crown and the short pieces that bend around the rear corners. The headliner is held in place by screws and by the support the lower pieces give to the large crown piece. In Ford trucks of the thirties to mid-forties, metal retaining strips are used above the doors to hold both the side headliner pieces and the weather stripping, which seals the door opening and is stapled or tacked in place before the headliner is installed. On the mid-fifties F-100 series, a trim retainer secures the panels around the door opening.

The cowl and door panels are held in place by screws or clips. On mid-fifties F-100–series trucks, a retainer located at the front door post helps hold the cowl panel in place. It's sometimes a little tricky to align the screw holes in the trim panels with the screw sockets in the body support structure. An easy

The finished seat is ready to be replaced in the truck. LeBaron Bonney Company

way to do this is to insert an ice pick into the hole and wiggle the point around until you locate the corresponding socket. Then you can move the panel so that the hole and socket align.

Door Handle and Window Crank Installation

After attaching the door panels, you'll want to replace the door handle and window crank. Small C-clips hold the door handle in place. A special tool, available from The Eastwood Company and trim supply shops, makes it easy to remove and replace this clip. Before replacing the door handle, it's a good idea to take a look at illustrations in a shop manual for your truck, or interior photos of correctly restored or original trucks, to check the correct orientation of the handle; usually it points up. Window cranks are typically held in place with screws.

Trim Installation

Trim items completing an interior restoration include armrests, sun visors, a firewall cover, and the glovebox. Until recent years, most pickups came equipped with only one armrest—on the driver's side. The passenger's-side armrest was an option. But passengers appreciate armrests too, and for trucks where these interior items are still available (from parts suppliers like Dennis Carpenter), you may want to add the right-door armrest.

Likewise trucks typically came with only one sun visor—again on the driver's side—with the passenger's-side visor an accessory. Since reproductions of the right-hand visor and bracket are available for many Ford pickup models, you may want to fit your truck with both visors.

The firewall cover, which is a padded trim piece

lining the passenger's side of the cowl below the instrument panel, is another item that should be replaced in a full restoration. Reproduction covers are available from Ford restoration parts suppliers. Note that these covers are held in place with special clips, which you should request when you order the cover. If you are concerned mainly with upgrading your truck's roadworthiness and appearance, chances are you won't replace the firewall cover unless the original is badly deteriorated or missing. If the firewall cover is good enough to leave in place, you may want to paint it to give it a new look.

Another interior item often needing replacing is the glovebox liner. Commonly, on an older truck, these cardboard liners are torn or missing. Replacement liners are also available for most years. Frequently the rubber glovebox door bumpers are also deteriorated or missing, and should be replaced.

Up until the Bonus Built series, in 1948, Ford sealed its pickup cabs against air drafts, with windlace that fastened on the inside of the door frame. This windlace, which looks like oversize fender welting, is placed against the door outline, with the round section projecting into the door opening and the flat tacking strip against the door outline. The windlace is held in place by tacks or staples that go into the tacking strip. The tack area is then dressed out with metal retainer strips.

Instrument Cluster Restoration

On most older trucks, the instrument panel will be faded, some of the gauges may have quit working, and the gauge faces will be faded and the markings obscured. As part of either a cosmetic upgrade or a complete restoration, you will want to refinish the dash. Services are available that do this work and can guarantee quality results. But you can also redo your truck's instrument cluster yourself.

The job of cosmetically refinishing an instrument cluster begins with disconnecting the wiring and removing the gauge set from the truck. If any of the gauges are not working, first do some electrical troubleshooting to determine if the problem is faulty wiring, a bad sending unit, or the gauge itself. (How to troubleshoot a nonworking electrical component is described in chapter 12.) Once you have figured

This Ford F-1 pickup is awaiting new interior coverings. When the old cardboard pieces have been removed is an ideal time to spray rust proofing into the kick panel cavities. LeBaron Bonney Company

New coverings give a truck's interior a nicely finished look. LeBaron Bonney Company

When some of the interior trim pieces are installed, the cardboard needs to be curved sharply. Here is a trick for doing this: Before bending the cardboard sections that fit against the rear corners of the cab, wet the back side of the cardboard with a mixture of 80 percent household ammonia and 20 percent water. The cardboard should be not soaked, just moistened. You will find that the damp cardboard will bend easily without crinkling. LeBaron Bonney Company

1948–1952 Ford Truck Gauge Restoration Tips
by Tim Master

When you remove the cluster from the dash, you will find that the outer ring has four little corners bent over. You can bend these back with a pair of pliers, allowing access to the gauges. Now remove the gauge housing from the chrome bezel. If this is the first time the cluster has been opened up, you might have to strike it with the palm of your hand to loosen it. Then you can remove the four gauges, followed by the speedometer dial, glass, and gasket.

I started the restoring process by sandblasting the housing. Because I couldn't find any information on whether the housings in these trucks were painted, I painted mine gloss-black (personal preference). You may want to send the bezel to a chrome shop to have it replated.

Since the gauges are extremely sensitive, you'll want to treat them with special care. To clean the gauge facings, I bought "canned air" from an electronics store, and a couple of number two artist's brushes. Use the air with short, quick blasts, and use the brush around the needle area so that you don't damage the needle itself. A wire brush cleans up the terminal ends for good electrical contact.

Each gauge has a calibration adjustment gear for the needle. But before attempting to adjust the gauge, check the sending unit where the gauge gets its signal. If you are getting a faulty gauge reading, usually the sending unit is the source of the problem.

To restore the gauges, prime the gauge dials and paint them off-white. On the 1951 and 1952 model year trucks, paint the center "bull's eye" silver and the surrounding black circle with a flat-black or antiglare black paint. On 1948–50 models, the dash gauges are arranged horizontally beside the speedometer. Their finish appears to be scalloped, and the color is closer to tan than white.

You'll probably need to cut a new cork gasket for the gauge assembly or assemblies (two pods were used for 1951 and 1952). Then reassemble the gauge units in the reverse order to that in which you took them apart.

out why the gauge or gauges aren't working, you can take the instrument panel out of the truck and remove the gauges. Refacing kits that you can apply yourself to give your gauges a like-new look are available for Ford trucks.

The instrument panel face will need to be repainted a flat (nonglossy) finish. Flattening agents are available from automotive paint suppliers. In many cases, the chrome bezel around the instrument cluster, found on Bonus Built and earlier-series trucks, will also need to be rechromed or replaced. Professionally rebuilt speedometers are available from vintage Ford suppliers for most years, or you can send your truck's speedometer to a specialty speedometer rebuilder listed in the Appendix. Typically the speedometer rebuilder will ask if you want the odometer to show the existing mileage or to be set to zeros.

Redoing your truck's instrument cluster makes an ideal cold weather project. The process requires more patience than skill, and an advantage of making this an off-season project is that any time delays, such as sending the speedometer off to the rebuilder, won't interfere with driving your truck.

Brightening up the gauge needles is an easy process. For many years, Ford used an off-white paint for these components. You should be able to find a close color match in the modeling paints at a hobby store. Either bottle or spray paint can be used. If you find the color match in a spray can, you'll need to spray some paint into the spray can lid or another suitable container. Then allow the paint to "flash off" (give it time to become thick enough to be

Speedometer Servicing
by Clarence Thielen

While the gauge unit is out of the truck, don't overlook servicing the speedometer. An oil hole and wick are located where the cable goes into the speedometer, and they never get oiled. That's why a lot of speedometers freeze up at about 85,000 miles from lack of lubrication. Also two brass worm gears in the speedometer should be greased.

If you have already had the dash out, or your truck doesn't need work on the gauge assemblies, the speedometer is removed with three screws that aren't hard to get at. You'll want to remember this lubrication tip and give the speedometer a few drops of oil periodically as the years go by.

brushable). For success, you will discover that the paint must be the right brushing viscosity. Next, using a very small, good-quality camel's hair brush, practice on a toothpick until you can complete the painting operation in one stroke. If you have to stop and start again along the needle, you will break the flow, and this break will be visible. After you have mastered painting the toothpick, you can graduate to painting the gauge needles. Be very careful not to drop any paint on the gauge face, as this will permanently mark the gauge and will spoil its appearance.

With the new interior in place, the truck is ready for an outing. LeBaron Bonney Company

Chapter 14

Rubber and Glass Replacement

As you set about restoring or upgrading your truck's appearance, it's easy to overlook the need to replace the rubber parts. Window and door seals, the rubber donut around the gas filler pipe, and the numerous other rubber items on the truck deteriorate at about the same rate as the rest of the vehicle, so the rubber's condition doesn't draw attention to itself—unless the truck has spent time in the southwestern United States, where the hot sun and dry air seem almost to fossilize the rubber.

If you're doing a ground-up restoration where all the rubber items will be removed, the need to replace these items will quickly become obvious because most will be destroyed as you take the vehicle apart. The time rubber parts sometimes get neglected is when you take the rebuilding approach. Here deteriorated window rubber or door weather seal may not be noticed until after the truck is painted. If this happens, the best time to remove and replace these rubber items has been missed. The smarter approach is to take inventory of the truck's rubber parts when you first plan the repairs and upgrades you intend to do on your truck, and to begin ordering replacements early in the restoration or rebuilding process.

Signs That Rubber Parts Need Replacing

Dryness, cracking, and missing chunks are not the only reasons for replacing a truck's rubber parts. Door weather seal can still look as if it's doing its job and have a spongy feel, but have lost enough of its resiliency so that air will hiss around the door opening when the truck is driven down the highway. As weather seal ages, the rubber loses its elasticity. When windshield rubber dries out and cracks, water starts to seep in around the glass. If the window rubber isn't replaced, you'll watch water drip into the cab when you drive your truck in the rain.

Usually rubber parts are replaced most easily as you are doing a related job; door weather seal, for example, is removed for painting and replaced as one of the finishing-up steps in the painting process. Besides the new rubber's doing its job of sealing out moisture and preventing squeaks and air noise, the

New rubber parts really help set off a truck's appearance. The ones in this interior cab view of a freshly restored 1961 Ford integral cab-box pickup include the windshield weather seal; vent window molding; and brake, clutch, and accelerator pedal pads.

benefit of renewing your truck's rubber parts is that the finished truck's like-new appearance won't be marred by unsightly dried-out or cracked rubber.

If you bought your truck from someone who has taken it partially or completely apart, or if it has already been through an amateur restoration, it may be missing many of the original rubber items. The problem in this case is figuring out what's missing. One way to learn what rubber items belong on the truck, and where they go, is to order the catalog of a vintage truck supplier like Dennis Carpenter and compare the rubber items shown in the catalog with those on your truck. Likely to be missing are draft seals for the clutch, brake, starter pedal, and floor shift lever; hood bumpers; grommets; and other small parts.

179

Rubber Parts Inventory

A convenient place to begin taking inventory of the rubber parts that need to be replaced on your truck is at the cab. If the door weather seal is original, it's probably cracked and large chunks may be missing. On newer trucks, this weather seal may be intact but should still be replaced. Also notice the condition of the large rubber grommet—often called a donut—around the gas tank filler pipe. Chances are the rubber has hardened, cracked, or torn, or is missing altogether. The rubber moldings around the windshield, backlight, and vent windows are also likely to be cracked and hardened.

Inside the cab, you'll probably find that the brake and clutch pedal pads and the boots for the accelerator, floor shift lever, and hand brake (on trucks where this brake is floor mounted) also need replacing. Even though they're likely to be missing, remember to add the clutch and brake draft seals that fit over the pedal shafts and rest against the floorboard (on trucks where the clutch and brake pedals depress through the floorboard) to your rubber parts inventory. Take a look at the condition of the steering post pad. You may need to put this item on your rubber parts list—and check for missing firewall grommets.

On the outside, check the condition, or the absence, of the hood and cowl lacing, door window channels, and glass sweepers. If your truck has had a lot of hard use, the lacing that forms a pad for the hood at the cowl—and at the radiator end on most of the older trucks—is probably worn and frayed and may even have been torn off and discarded. To inspect the condition of the lacing, just open the hood and look for a thin strip of woven fabric running up the sides and over the top of the cowl and nose assembly. In the door and vent window channels, "whisker" lining that keeps the glass from rattling is probably badly worn and may be missing. On most older trucks, the sweepers at the bottom of the door window openings also need to be replaced. These attach to the door at the base of the window frame and keep water from dripping down inside the door. They also clean the window as it is rolled up and down.

As you walk around the truck, looking carefully at the condition of rubber parts, you may notice rubber gaskets around the edges of the headlight bezels. These are likely to be cracked and should be added to the rubber parts list. For trucks with a cowl vent, add this seal to the rubber parts inventory. On most older pickups, fender welt is used to seal the seam between the rear fenders and the box. Welting was also used as a cowl-to-fender seal through 1956.

At the truck's business end, you may notice that two important rubber parts—the tailgate chain covers—are missing. These should be added to the inventory. If you trace the path of the wiring harness, you'll probably discover that grommets intended to protect the wiring harness as it passes through holes in the body are also missing. Replacement grommets are not supplied with new wiring harnesses, so these too go on the list.

Now crawl under the truck and inspect the shock absorber bushings. If the truck predates airplane-style telescoping shocks, the lever arm shocks will probably be removed for cleaning, refilling, and possibly repair. When they're replaced, you'll also want to install new rubber bushings on the shock arms.

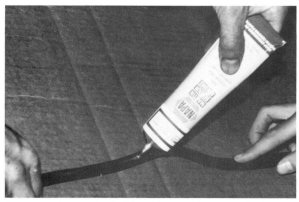

Door weather seal is held in place with a special adhesive available from auto parts stores.

Weather seal adhesive comes in two types. One type is applied only to the metal, and the rubber is pressed in place before the glue sets. The other type acts like contact cement. With this variety, a bead of adhesive is applied to both the metal and the rubber.

Back out in the light again, did you notice the condition of the door bumpers, on trucks so equipped, when you looked inside the cab? And on thirties-era trucks, have you checked the condition of the hood pads at the corners of the hood, or are they missing altogether?

Rubber parts differ with various models of trucks. You may notice rubber items not mentioned here. In any event, make the list as complete possible; then you can go shopping.

Replacement Rubber Sources

For mid-sixties and later models, the place to begin looking for rubber parts for your truck is your Ford dealer. Many collectors don't think of a new-car or -truck dealer as a parts source because they believe that manufacturers only stock parts for the latest models, or that a manufacturer's prices will be higher than those of a specialty parts source.

Typically, neither is the case. The dealership can usually order most parts for trucks built within the last decade and many parts for earlier trucks. If you find a dealer who has been on the same site for several decades, you also may run onto a sizable parts cache for older models. As far as the price of these parts goes, a dealer's retail prices are often lower than those of a specialty supplier, and a dealer's wholesale prices are likely to be substantially lower.

Specialty Ford parts suppliers like Dennis Carpenter are the next source to check for the rubber items on your list; several are listed in the Appendix. Two prime sources of rubber items are Metro Moulded Parts and Steele Quality Reproduction Rubber.

Metro Moulded and Steele Rubber carry rubber parts for a wide range of vintage cars and trucks. If their catalogs don't list exact items for your truck, take measurements—or in the case of weather stripping, identify the profile of the original—and compare them with those of the "universal" items shown in the catalog. The same rubber parts are often used on many vehicles.

Although universal rubber may work for door seal, don't fall into the trap of ordering a universal floor mat. Reproduction floor mats meeting OEM standards are available for most years of Ford trucks at reasonable prices. Although it's usually possible to trim a universal floor mat to match your truck's needs (they're molded oversize to fit the widest possible range of vehicles), problems arise in trying to match any precut holes with the actual location of the clutch and brake pedals, floor shift lever, accelerator, and starter control on your truck. In addition, universal mats usually aren't contoured to your truck's transmission hump, so they won't lie flat.

If a reproduction floor mat isn't available, it's always possible to have a trim shop cut carpeting to

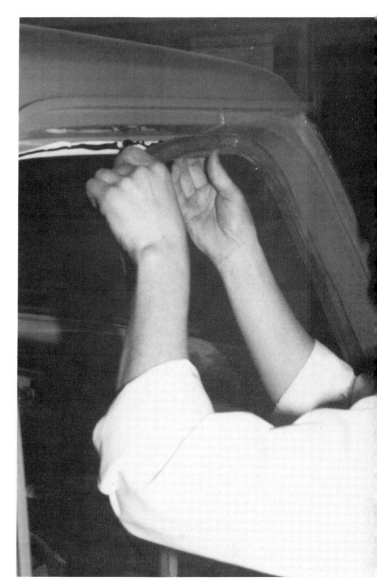

If you follow the instructions with the adhesive, when you press the weather seal in place, a tacky bond forms almost immediately.

fit the cab floor. Carpeting has two drawbacks, however. First, no one put carpet in a truck until the mid-seventies, so carpeting in a pre-seventies truck definitely isn't original. Second, if you intend to use your truck for hauling a boat to the lake and toting occasional supplies home from the lumberyard—not especially dirty work, but jobs where you will track some grit inside the cab—you'll find it easier to sweep a floor mat than to vacuum a carpet.

Occasionally it's possible to find original rubber parts at swap meets. To be sure you've found the right part, look up the part number in a parts book beforehand or, better yet, bring along a cross section or the complete original part with you. If you buy any new old-stock (NOS) rubber parts, make sure they are still pliable, and don't listen to a vender's

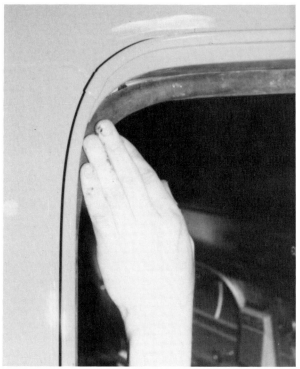

It is important to press the weather seal tightly into contours.

story that a stiff rubber part can be softened with a couple of coatings of Armor All. Rubber is like toast; once it has turned crisp, it can't be softened again.

On a light truck that's twenty or more years old—especially if the vehicle has spent most of its life in warm, sunny climates—you will probably find it necessary to replace most, if not all, of the rubber parts. This will typically cost between $500 and $700. It's a good idea to gather the complete rubber inventory early so that you will be able to install new rubber items whenever they're called for in the rebuilding process.

The guideline here is to purchase replacement rubber only from parts venders who carry quality parts: a Ford dealer, reputable restoration supplier such as Dennis Carpenter, or specialty rubber parts manufacturer such as Lynn Steele and Metro Moulded. Avoid dealing with discount mail-order parts outlets that only carry parts for older cars and trucks as a sideline. Although their catalogs sometimes show a lower price for the few rubber parts they list, the discount parts are likely to be of inferior quality and, worse yet, may not fit.

Rubber Parts Installation

Rubber parts are typically installed as part of a larger process (like replacing the windshield) or as a

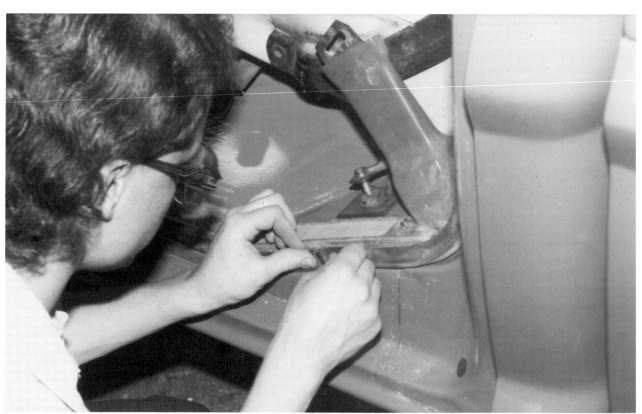

As part of your research, either before you take your truck apart or from an original or correctly restored truck, note the location of the door weather seal seam. Then, when you fit the weather seal, you can place the seam in the same location as the factory did.

detailing step after the exterior has been repainted or the interior has been refinished. Usually the only difficulty is making sure these parts are not forgotten and, in some cases, recalling exactly where they go. This is why it is important to take photos and make notes on the locations of rubber parts before taking the truck apart. If the truck is already apart, you'll have to find out where the rubber parts go by taking photos and noting the locations of the rubber items on other original or restored trucks you may find in scrap yards or see at club events. Catalogs from vintage Ford truck parts suppliers like Dennis Carpenter and the body maintenance and repair section of the Ford shop manual for your truck are good references on the proper location of door weather seal and other rubber items.

In almost every case, rubber parts are installed after painting. This is the sequence in which they were originally placed on the truck. As a result, the rubber items left the factory unpainted. Hood lacing should also be attached after painting, and the same applies to door window channel and sweepers.

Fender welting is also replaced as part of an assembly step. This means that the fenders should be painted separately rather than on the truck. If you do paint the fenders on the truck, loosen them and replace the welting after painting. If you remove and paint the fenders, you'll find that great care is required to keep from nicking or scratching the paint as you bolt them back into place. On later-model trucks that don't use welting to fill the seam between the rear fenders and the box, a bead of liquid antisqueak or rubber caulk sealer should be applied to the fender lip where it contacts the box, to prevent moisture from penetrating this seam.

Rubber parts that are part of a mechanical repair, such as shock absorber bushings, are installed during the repair process. If you're replacing the floor seal on the steering column, it'll be necessary to remove the steering column from the truck and disassemble the tube from the steering box. This effort may be worth it on a frame-up restoration where you may be overhauling the steering box and giving the column a high-gloss finish, but if the steering box only needs adjusting and you plan to repaint the column while it's still in the truck, taking the steering assembly apart just to replace the post pad isn't worth the trouble. If the original seal is badly deteriorated or missing, you can slit a replacement pad, fit it onto the steering column, and glue the seam. It'll take a careful look to recognize that the seal wasn't installed the way they did it at the factory.

Many rubber items are replaced as part of detailing after the truck is painted. Examples inside the cab include the pedal pads, draft seals, and glovebox bumpers; examples on the exterior of the truck are the gas tank filler pipe grommet and tailgate chain rubbers. Most of these items just slip

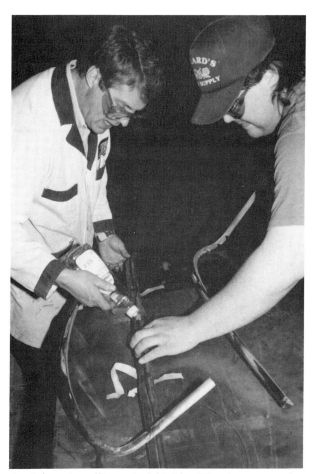

A trick to installing windshield rubber is to coat the grooves for the glass and brightmetal trim with dish soap. The soap acts like a lubricant and makes it easier to fit the glass and the trim into the weather seal channels.

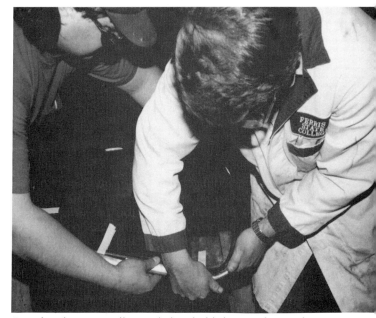

Four hands are usually needed to hold the weather seal onto the glass while fitting the brightmetal trim.

The weather seal is held onto the glass with strips of masking tape.

into place. If the fit is snug, as will most likely be the case with the filler pipe grommet and tailgate chain rubbers, a light coating of liquid dish soap on the rubber surface will allow the rubber parts to slide easily in place.

Door Weather Seal and Cowl Vent Seal Replacement

Door weather seal attaches with special glue. This process is relatively simple but does involve several steps. At some point, the old rubber has to be removed. Usually this is done in preparation for rust repair and repainting. Although the old weather seal can often be pulled loose by hand, usually some of the rubber and glue stick to the metal and have to be scraped clean with a putty knife. If you just want to replace the rubber and aren't planning to repaint, you will need to scrape very cautiously to avoid scratching the paint. Prep solvent, used to remove wax and tar from a finish in preparation for painting, can be used to soften the old glue and should be used to clean the surface before gluing new weather seal in place.

The metal around the old weather seal may look solid, but you're likely to find rust scale hiding

After the windshield is replaced, any gaps between the weather seal and the window opening are filled with rubber caulking. If you have never replaced a windshield, the job of installing this large and easily cracked glass piece is best left to professionals.

behind hardened or decayed rubber. If so, it's essential to clean away the rust before proceeding. When no rustout is present, you can remove surface rust with a wire brush or spot sandblaster, then treat the metal with naval jelly, a phosphoric acid jell available in hardware stores. The gray residue left by the jell should be wiped off with a damp cloth before painting. The metal can then be primed and touchup painted. If rust damage is extensive, metalwork and refinishing will be required.

Weather seal cement is available at auto parts stores. This product comes in two types. One is applied only to the metal, and the rubber is pressed in place before the glue sets. The other acts like contact cement. With this type, a bead of glue is applied to both the metal and the rubber. When the glue on both surfaces becomes tacky, you press the weather seal in place. A strong bond forms almost immediately, making it difficult to pull the weather seal back off if it isn't positioned just right the first time.

With either type glue, it's important to spread only enough to hold the weather seal. Any extra can be cleaned up with prep solvent, but it doesn't come off easily. Sometimes the rubber wants to fall loose on the bottom of the door. It can be held in place with strips of masking tape while the glue dries. Normally the glue is sticky enough to hold the rubber snugly in place without the tape. Be sure to give the glue time to set before closing the doors.

Cowl vent seals are attached using the same method.

Vent Pane and Door Glass Weather Seal Replacement

Several weather seal and glass antirattle items need to be replaced to keep rain from entering the truck cab, prevent air from hissing around the door glass, and give the door glass area of your truck a finished look. Ford trucks prior to the 1953 F-100 series did not use vent windows, so only the door glass seal (called the glass run or window channel) and glass antirattlers (also called window sweepers) are replaced on these trucks. From 1953 models on, the vent window rubber and the division bar that forms the forward run for the door glass are also replaced. These parts are available from Dennis Carpenter and other vintage Ford truck parts suppliers.

Replacing door glass weather seal requires

Replacing door glass weather seal requires removing the door glass and the vent window pane.

To remove both the door glass and the vent pane, it will be necessary to loosen the window division bar, which is held in place by screws that are reached by removing the cover plate on the inside lower portion of the door.

removing the door glass and vent windowpane, on trucks so equipped. This glass doesn't have to be removed for repainting, and usually isn't unless the truck is being restored—in which case, the doors have probably been taken off the truck, completely disassembled, and sandblasted or chemically stripped.

To remove the glass, you will take off the cover plate on the inside lower portion of the door; remove the screws holding the division bar in place, on trucks with vent windows; and then lower the door glass and disconnect the glass from the regulator arms. You'll find it helpful to remove the window regulator mechanism too because having this item out of the way makes removing the window division bar easier. The regulator mechanism just unscrews from the inner door panel. Don't forget to take off the window crank. On Ford trucks through 1952 without vent windows, this step can be skipped.

Now you can remove the glass sweepers located at the bottom of the window opening. The sweepers are held in place with clips and are pried loose with a screwdriver. If you're doing this after painting the truck, run a couple of layers of wide masking tape

If you are replacing the vent window weather seal, it will be necessary to remove the vent pane from its frame assembly. This may require driving the swivel post out of the frame assembly. Note that the nut has been threaded onto the end of the swivel post, to prevent the hammer force needed to free the vent window from damaging the threads. Be sure to hold onto the vent window while driving it out of the frame, to keep it from falling and breaking.

Before installation, the new weather seal is wiped with wax and grease remover to give the adhesive a clean surface for strong bonding.

around the edge of the window to keep from scratching the paint. Usually the clips will break off. You can pull them out of their socket with needle-nose pliers.

The door glass run comes out next. It too is secured with clips. On many models, including the 1953 and later F-100s, the glass run also fastens with screws to the lock end of the door.

On trucks with a vent window and a division bar, these items are the last to be removed. The vent window is held in place by screws in the frame assembly. On early F-100 trucks, these screws are at the top and bottom of the frame near the division bar. The bodies and cabs section of your shop manual will show the location of these screws, and you can see them by looking carefully in the rubber window seal channel. As soon as the division bar is pulled back, the vent window and its frame will pull free. Since the screws securing the division bar have already been removed, this bar can now be pulled back to remove the vent window assembly. Since a new division bar will also be installed, it's necessary to pull the bar out through the access panel at the bottom of the door. It may take some maneuvering and twisting, but the bar will come out.

With all the window parts out of the door, now is the time to replace the old glass, if necessary (this process is discussed in a later section of this chapter) and clean up the vent window assembly. If the vent glass frame needs refinishing (this frame is painted black), and the glass is not being replaced, mask the glass and paint the frame.

Everything, including the new window seal, goes back together in the reverse order from that in which it came apart, but first you need to install new rubber weather seal in the vent window frame. The rubber presses into the frame channel, but to make a good seal and ensure that the rubber stays in place, it's a good idea to lay some weather seal glue in the channel first. Be sure to line up the screw holes and pivot point indentations in the rubber with their corresponding locations on the frame. You will set this part aside for a few minutes while you reinstall other items.

Now you can form and install the door glass run. The new run comes as a straight piece, so it has to be bent to follow the window contour. The best way to do this is to use the door glass—already removed from the truck—as a form. Before installing the run, place the glass back inside the door. Now you can align the clips that hold the run in place in their holes in the door and press the run in place. Remember to replace the screws at the bottom of the run. These screws thread in from the lock end of the door.

The new division bar goes inside the door next. Then you can reconnect the door glass to the window regulator. If the truck you're working on has vent windows, the vent window frame can now be screwed back in place. Be careful not to scratch the paint as you position the vent window frame in the door. Next line up the screw holes in the vent window frame with their attaching points in the door, and insert and tighten the screws. With the vent window in place, you can swing the divider bar into location against the front of the window run and install the attaching screws.

All that remains is to replace the glass antirattlers at the bottom of the window opening. When doing this, make sure the door glass is rolled down as far as it will go. You risk breaking the glass if you press

Weather seal adhesive is then applied to the rubber.

against it while pushing in the clips that hold the antirattlers. The nontempered automotive glass found in older trucks cracks easily. If the glass is in the way as you are installing the antirattlers, it's a good idea to detach the glass frame from the regulator mechanism and let the glass rest against the bottom of the door.

Some adjustments to the positioning of the division bar may be needed to make sure the vent window closes properly and the door window glass rolls up and down easily. Spraying WD-40 onto the regulator mechanism and the slide area on the bottom of the door glass channel will also reduce the effort needed to roll up and lower the glass. You'll find as well that a coating of WD-40 on the new vent window rubber will allow that window to open and close more easily; the new rubber tends to "cling" to the vent pane frame. When the glass moves easily and seals properly, you can replace the access cover in the bottom of the door, and this job is done.

Windshield and Rear Window Glass Seal Replacement

If the truck is receiving a complete repaint, the windshield and rear window glass are usually removed. This allows the metal underneath the window rubber to be painted and gives an opportunity to replace the weather seal. Also, if the windshield is scratched or fogged, it's time to look for new glass. Putting these pieces back in the truck is a job best left to professionals. It's not that replacing windshield and rear window glass is all that difficult, but these are large glass items that can be broken if handled improperly—and they're costly to replace. So having a glass shop replace them is the best bet. The professionals are also better able to make sure the new rubber is sealed to prevent water seepage.

Rubber Parts Preservation

Sunlight causes rubber to dull and harden. To preserve the glossy look of new rubber and to keep rubber parts soft and pliable, they should be given a coating of rubber moistener and preservative each time you wash your truck. The product to use is Armor All, and it's available at any discount mart's automotive department. Armor All can be sprayed directly onto the rubber, or applied to a rag and wiped on. The Armor All treatment restores that

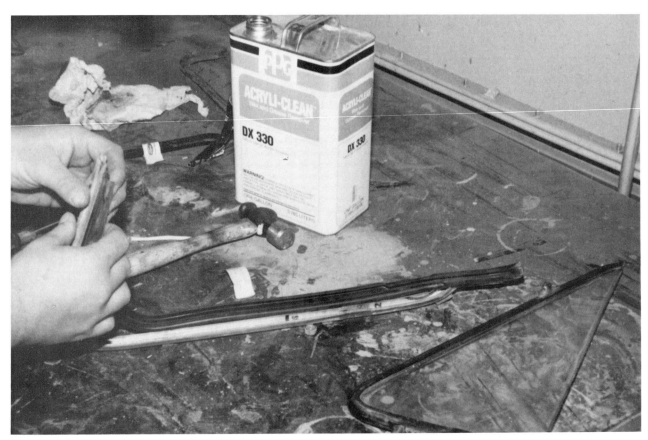

The new weather seal presses tightly in place. Here new vent window rubber is being fitted into the frame assembly.

shiny black look and keeps the rubber from drying out and cracking.

New rubber and weather seal change a truck's appearance more than you may realize until you do a before and after comparison. Fresh rubber parts give a finished look while making the truck more comfortable to drive by sealing the cab against air drafts and water leakage.

Door and Vent Pane Glass Replacement

Broken or cracked window glass obviously needs replacing, but you may decide to replace your Ford truck's windshield, rear window, or door glass for many other reasons. With age and exposure to sunlight, automotive safety glass turns cloudy along the edges. This cloudy band can eventually expand to the point where vision is obscured. Although it's recommended that the hobbyist restorer have the large glass pieces for the windshield and rear window replaced by a professional glass shop, replacing door glass carries little risk of breakage and is a job you may want to do yourself.

The closest source of replacement glass is the local glass shop—not a hardware store, but a business that specializes in replacing commercial windows and auto glass. This shop will use the old glass as a pattern to cut new replacements, so if the glass is missing or broken, you may have to use another source such as a reliable restoration parts supplier.

The major problem with glass from a local shop is that the new door glass will lack the factory markings. These markings, which include the manufacturer's name or logo and sometimes the production date, are etched into factory glass, and vehicles entering rigorous Antique Automobile Club of America (AACA) or another national competition can lose points if their glass doesn't have them. If details like the markings on the glass are important, you will need to purchase replacement glass from a glass supplier like GLASCO, which can etch the original factory markings into the replacement.

It is sometimes possible to find NOS glass at swap meets, but if you find replacement glass this way, be sure to check the correct stock number—taken from either the original glass or a parts book—against the one on the replacement glass. It's hard to know just by looking at it whether a piece of glass

Next the vent window is replaced in the frame.

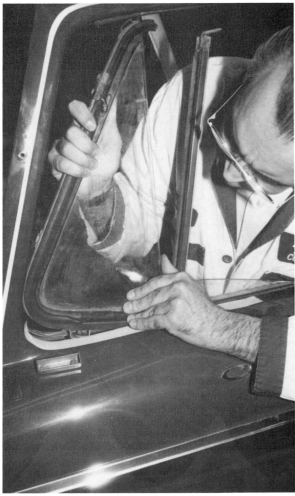

Then the entire vent window assembly is installed back in the door. If you are replacing the weather seal on both the door glass and the vent window, the vent window and the division bar are the last items to be removed and the first to be replaced.

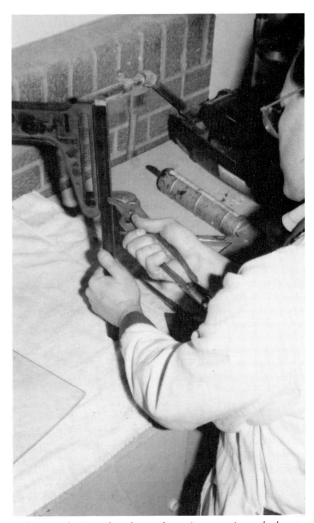

When replacing the door glass, it sometimes helps to pinch the metal channel on the glass support bracket slightly closed, to make sure the glass fits snugly into the channel.

that is represented as authentic for your truck is in fact correct. Also carefully inspect the edges of the glass to make sure the seal between the glass layers has not deteriorated. If this seal has broken, you may notice clouding around the edges of the glass.

In preparation for replacing the door glass, you'll need to gather several feet of channel liner (a thin, rubbery strip available from most body shops that do glass replacement) and a tube of windshield sealer-glue.

When you have all the new glass and supplies, and have cleaned and painted the metal channel that the glass fits into, you're ready to install the glass. The first step is to cut a length of channel liner slightly longer than the metal channel. To help ensure that the glass doesn't work loose at some point in the future, it's a good idea to spread a bead of glass adhesive along the bottom of the channel before fitting the glass. Some adhesive will squeeze out of the channel when the glass is inserted, but this excess can be scraped off the glass with a putty knife and cleaned up with prep solvent (a product used to prepare a surface for painting).

Now wrap the length of channel liner over the edge of the glass that will be inserted into the channel, hold the liner in place, and press the glass into the channel. If the fit is tight, place a hardwood board on the channel and rest the glass on another board. Tap very gently on the wood piece that you've placed against the channel; hard pounding can crack the glass. Fitting door glass into its metal channel is usually not very difficult. When the glass is firmly seated in the channel, you can cut off the excess liner and clean up the glue.

Replacing vent pane glass follows the same process. Be very careful not to exert any sideward pressure. Even a very slight amount of pressure against the glass's flat surface can crack the glass.

Chapter 15

Box Restoration

One job that is nearly always required in restoring an old truck is replacing the wooden flooring in the pickup box. The wood may be still intact, but scarred and scraped—a condition that gives a truck a well-worked look. Whether you are restoring or rebuilding, you'll want to renew your truck's wooden pickup box flooring. This is a job you can do yourself; all the parts and pieces you'll need are available from a number of suppliers. When you're done, you'll not only have the pride of your accomplishment, but you'll realize that nothing sets a vintage truck off like new wood in the pickup box.

Restoration of wooden bed floors in trucks that are from the twenties to sixties can be accomplished in several ways, but when it comes to winning show trophies, care and professional quality are all-important. In this section, cabinetmaker Bruce Horkey shows the step-by-step procedure he uses to restore wooden pickup beds. Horkey also makes sideboards and custom pickup bed parts.

(Unless otherwise indicated, all photos in this sequence are courtesy of Bruce Horkey Cabinetry, RR4, Box 188, Windom, MN 56101.)

Wooden Flooring Replacement
by Bruce Horkey

Step One
Begin by cleaning up the box and floor. Take pictures of all the details, and make notes of all particulars pertaining to fit and sequence. List all fasteners, noting size, type, and location. Remove the fenders, lights, and wiring. Order new wood, skid strips, and angle braces. If the box sides, front, or tailgate are badly dented or rusted and need replacing, also order these parts. New metal is available for Ford pickups from the twenties through the sixties.

Step Two
To make room for working underneath, remove the pickup box from the frame and set it up on sawhorses or stands. Remove the old bed wood, skid strips, bolts, and cross-members.

Step Three
Strip or sandblast the box metal, do any needed dent and metal repair work, and refinish the box sides, front, and tailgate (see the "Pickup Box Restoration" section later in this chapter). Square up the box and snug up the bolts on the front panel and the rear main cross sill. Varnish or paint the new wood (see the "Wooden Bed Varnishing" section later in this chapter).

Step Four
Clean the angle pieces along the sides of the bed, and the area around them. These will be the first items replaced. Scribe a reference line above and below the angles on the box sides for replacement reference. Measure from the bottom of each box side to the underside on the angle. Write down this measurement. Make notes on hole locations and angle location, on the box side.

Step Five
Center punch the spot welds attaching the angle to the box. These welds will be drilled out with a spot weld cutter. Set the depth of the cutter to drill only through the angle. After drilling out the spot welds, remove the old angle.

Step Six
Set the new angles in place, following the notes you made earlier, and mark the hole locations on the box side. If you are working with angles that have no holes on the shorter side, note the fender and stake pocket locations and decide where to drill holes to attach the angles to the box sides. Drill the holes in the box side, and attach the angles; for originality, the angles should be spot welded in place instead of bolted.

Step Seven
Set the edge boards in place and mark the holes to attach the angles to the edge boards. Remove the boards and drill the holes with a 3/8in-diameter bit.

Step Eight
Install the edge boards and drop the bolts

Step 1

Step 2

Step 3

Step 4

Step 5

Step 6

Step 7

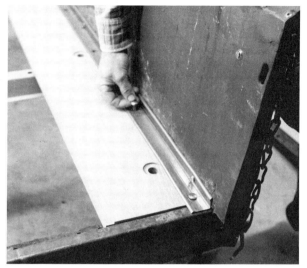
Step 8

Step 9

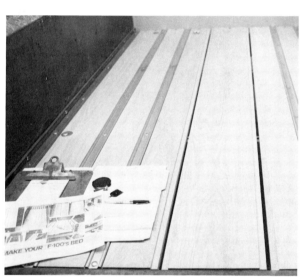
Step 10

Step 11

Step 12

through the angles. Loosely install the crossmembers.

Step Nine
Install one board and skid strip at a time. Insert the bolts, nuts, and washers loosely from the inside of the box by reaching over and under the boards. Attach the boards in an alternating pattern—first on one side of the box, then on the other.

Step Ten
Install the center board and the last two skid strips and bolts. Measure the box diagonally to ensure squareness. Align the boards, and adjust everything to fit.

Step Eleven
From the underside, loosely install the remaining nuts and washers. Make one final check for alignment and squareness. Hand tighten all nuts. After you're satisfied with the fit, set the box on the truck frame. Place a blanket over the front box panel to protect it and the cab from bumping together and scratching the finish.

Step Twelve
Drop in the hold-down bolts and washers. Square the box on the frame and make alignment adjustments, then bolt the box in place.

Caution: Don't overtighten the skid strip bolts. Now stand back and admire your work.

Wooden Bed Varnishing
by Bruce Horkey

Although Ford pickups didn't leave the factory with varnished flooring on the pickup box, most truck owners prefer a high-gloss finish on the pickup planking, over the dull black or painted coatings that were applied as the box came down the assembly line. Actually the varnish finish is important for more than just appearance. It also protects, and in so doing, extends the life of your truck's bed wood. Here are the steps to follow in applying a quality varnish finish.

Although Ford pickups didn't leave the factory with varnished flooring in the pickup box, most truck owners prefer a high-gloss finish over the dull black or painted coatings that were applied on the assembly line. Bruce Horkey

Tools

Before you begin, you will need to gather the following tools and materials:
- file with a safe edge (no teeth on one side)
- 120- or 280-grit sandpaper
- tack cloth
- Scotch-Brite pad
- 2 varnish brushes
- 2 2lb coffee cans or similar containers
- 2qt marine spar varnish
- 1qt quality paint thinner
- wipes
- sawhorses

Varnishing Procedure

The first step is to take the file and stroke off the sharp edges of the boards. This will prevent the edges from snagging the paintbrush and ensure better adhesion for the varnish. When filing the edge of the groove, lay the file on the groove at a 45-degree angle and allow the safe edge (the edge with no teeth) to ride on the bottom of the groove.

Next sand the faces and edges of the boards by hand with 120-grit sandpaper. When you are satisfied with the smoothness of the boards, you can prepare for varnishing.

The space you select to do the varnishing should have good ventilation; this is healthier for you and gives faster drying. Clean the area thoroughly. Sweep the floor well, then wet it down to settle the dust. Now arrange the boards on sawhorses and get out the varnishing supplies.

To make sure no dust is on the boards, vacuum all surfaces of the wood, then wipe the boards with a tack cloth.

For the first varnish coat, mix 1/2qt of varnish with 1/2qt of thinner. A 2lb coffee can makes a good mixing container. Apply this thinned varnish to the wood completely and evenly, coating all surfaces. Let this first coat dry overnight—8 to 12 hours. When the first coat is dry, move the wood away from the varnishing area and lightly sand with 280-grit sandpaper. Wipe the boards again with a tack cloth.

A furniture-like finish to the pickup bed floor really gives the truck a well-cared-for look. Noel Glucksman

Bring the wood back to the varnishing area and apply a second coat of fifty-fifty thinned varnish. This second coat will ensure that the varnish penetrates thoroughly into the wood. After curing, these first coats provide a barrier that will let subsequent layers build up on the surface. When the second varnish coat has dried thoroughly, again move the wood away from the varnishing area and lightly sand with 280-grit paper. Wipe the boards with a tack cloth to remove all dust.

The third coat is thinned only 25 percent; it consists of 75 percent varnish and 25 percent thinner. This coat is also sanded with 280-grit sandpaper. Be careful not to cut through the varnish on the edges of the wood. Tack off the boards to remove all dust.

For the fourth coat, use a fresh can of varnish and a fresh, clean brush. Do not shake or stir the varnish. Bubbles can present a problem and are sometimes created by wiping off the brush on the

---

Holey Floorboards, Bedman!
by Joel Miller

Something drastic had to be done. When I looked into the back end of my 1958 Styleside, I could see more garage floor than pickup floor! Until I came to its rescue a few years ago, the truck obviously had spent most of its life in the great outdoors. The bed wood was so rotten, I pulled most of it out by hand. And several bolts were so badly rusted, even penetrating oil wouldn't help.

How to get all those old bolts out? A hacksaw or nut splitter would be much too slow. So I got serious. I held the bolts with a ViseGrip pliers, and cut them off with an angle grinder. That worked like a charm.

I had some nice Douglas fir boards on hand to use for a new floor, but those rusty skid strips had to go. At this point, I turned to Bruce ("Bedman") Horkey. Horkey specializes in replacement bed parts. He has liked old trucks for years. During the restoration of his own pickups, he was looking for a source of parts. Since he already was pursuing a woodworking career, he decided to start his own company that combined his love of pickups and carpentry. Horkey's catalog has a fine selection of precut oak boards, primed or stainless skid strips, mounting hardware, and more, for pickups of all popular makes and years. You'll find box fronts, sides, and tailgates; fenders; tonneau covers; varnish; floor conversion kits; running boards; and custom oak panels for the box exterior.

While waiting to receive Horkey's shipment, I went to work getting the new floorboards ready. I cut all the boards to the proper length and width on the table saw. Then, with a saber saw, I notched the two outer boards to clear the inside of the fenderwells. By using a salvaged chunk of the original floor, I determined how wide and deep to cut the slots the skid strips fit into. Out came the router, and the new boards got their edges grooved.

The next step was to set both outer boards in place and mark where bolt holes should be bored along the sides of the box and for the main hold-down bolts. To avoid splitting the wood and to get clean, properly centered holes, I used good-quality wood-boring bits. That worked much better than trying to use ordinary drill bits. It's amazing how the need for special tools multiplies when you're trying to do a professional-looking job!

Next I used a belt sander to give the boards a smooth finish and then brushed on three coats of clear, gloss liquid plastic. The boards look beautiful with their warm amber glow and excellent grain patterns.

Meanwhile my order from Horkey had arrived. The end of the tunnel was nearing. I was itching to use the new sandblaster my wife bought me for Christmas. I loaded the hopper; put on my respirator, hood, and gloves; and fired up the compressor. I looked as if I was about to handle radioactive waste with that space suit on!

In about an hour, I had completely stripped the box around the perimeter where the new floor would be mounted. The driveway looked like a beach, but the sandblaster did a wonderful job removing the old paint and rust. I blew all the loose sand off the truck and applied primer and a couple of coats of enamel to the newly bared metal.

Now it was time to put in the new floor. One difficulty I found was trying to place flat washers, lock washers, and nuts separately onto the ends of several bolts that were protruding into areas I could hardly get my fingers into. I solved that at the workbench by making up several nut and washer "packs" ahead of time. I placed the hardware onto the bolts and finger tightened the nuts. Then I tightly wrapped short pieces of electrical tape around the washers and nuts. When I took the nuts off, the washers went along for the ride, and I could screw the packs on with two fingers in the tight areas. The plastic tape then just pulled off.

After the new boards and skid strips were installed, and all the bolts had washers and nuts on them, I crawled around under the bed and tightened each fastener. The last step was to jack up the floor support cross beams slightly at the locations of the eight hold-down bolts. Webbed insulators and thin wood blocks were then slipped between the beams and the truck's frame. After those bolts were tightened down, the new floor was done.

It looked so good compared with the rest of the truck, I knew I had to get busy with more restoration work!

Replacement box sheet metal, as well as entire replacement beds, is available for most Ford trucks from the Model A and T era into the sixties.

This F-1 undergoing restoration is wearing an entire new reproduction bed and rear fenders.

edge of the can. To prevent bubbles, wipe off the brush as necessary on the edge of an empty coffee can. This fourth coat is not thinned. Apply it first to the back side, edges, and ends of the boards—and allow the boards to dry. Then apply the varnish full strength to the face side. When the boards have dried thoroughly, they can be assembled in the box.

When you have expended this much care and labor to achieve a furniturelike finish on your truck's pickup box floor, you will want to treat the wood like fine furniture. Use a soft cloth to wipe off any dust that accumulates. A tonneau cover is recommended to protect the box flooring from sun and weather.

Pickup Box Restoration

Rust and dent damage to the pickup box is very common. Depending on the region where the truck has spent most of its working life, rust can vary from surface corrosion where the paint has worn thin, to complete rustout of the side panels, cross braces, stake supports, and tailgate.

Fortunately replacement sheet metal and entire replacement beds are available for most Ford trucks from the Model A– and T–era into the sixties. Note that these replacements are for trucks with the narrow Flareside beds. Replacement sheet metal for the wide Styleside boxes that first appeared in 1957 is more difficult to locate. Also, Styleside boxes, with their welded construction, are more complicated to repair. The integral cab-box design used on 1961–63 Ford pickups presents even greater repair difficulty if the box is rusted. Original replacement metal for these vehicles is very hard to find and expensive. Usually if a truck with this construction is badly rusted, you'll probably end up using it for parts and finding another with better sheet metal.

Box restoration begins by determining the extent of the rust damage. This is done by jabbing a screwdriver into suspected rust areas, as described in chapter 10. The stake supports are particularly rust prone, as is the metal covering that was used over a wooden subfloor on Ford pickups from the 1930s to 1950. If the fender welting from the seam between the rear fenders and box sides has rotted out or been removed, it is also possible that the box sides are

If you are building up a new box from replacement parts, Bruce Horkey recommends laying out the parts on sheets of plywood to prevent the metal from being scraped on the garage or shop floor. Bruce Horkey

With the sides and ends riveted or bolted together, the new box is taking shape. Bruce Horkey

The support braces attach next. With these in place, you're ready to install the flooring. Bruce Horkey

A tonneau cover is recommended to protect a cabinet-quality box floor from rain and sun. Bruce Horkey

rusted in this area. If the box sides are sound and not badly dented, you may decide to replace only the stake supports. These can be removed by drilling out the spot welds, purchasing new stake supports from Mack Products or another vintage Ford truck parts vender, and welding the new supports onto the side panel.

The front and side panels, cross-members, tailgate—all the parts needed to completely rebuild the narrow Flareside box—are available for 1928–72 Ford pickups. Ford used rivets to assemble its pickup boxes, so to replace a panel, you'll need to drill out the old rivets, then align the new panel and clamp it in, and replace the rivets. Rivets of the correct dimension and size are available, as are rivet-setting tools that form the correct shape head. It's not necessary to heat the rivets. To install replacement rivets, insert them in the original hole, have a friend hold a hardwood block against the rivet head, and then place a rivet-setting tool against the small end and strike the tool with a hammer. Usually it takes several hammer blows to flatten the rivet enough so that a tight joint is made. The metal flooring that Ford installed over the wooden subfloor in pickup boxes through the thirties and forties is also available, so this item is also easily replaced.

The pickup box is likely to be the most beat-up part of the truck, but all the parts for Ford narrow-style boxes, from sheet metal to flooring, are being manufactured. No metalworking skills are required to replace parts or completely build up a pickup box, so this is a project that virtually anyone can do.

A wooden box liner will protect a refinished steel floor or new wood flooring and allow you to carry cargo in your restored pickup. Jim Baker

Chapter 16

Brightmetal Restoration and Replating

One advantage of restoring a truck, as opposed to a car, is that the vehicle has only a small amount of brightwork (chrome and stainless trim) you'll need to redo. Of course, exceptions exist, most notably the late-fifties Deluxe Cab models with their chrome bumpers and grille and bright moldings around the windows. On most trucks, however, chrome trim is limited to the headlight rims, possibly the grille and front bumper, nameplates, door handles, and miscellaneous small knobs and control handles inside the cab. In the sixties, stainless and aluminum replaced chrome on the grille, again simplifying this phase of the restoration process.

Nothing brightens up an older truck like gleaming chrome and shining brightmetal trim. In contrast, neglecting the trim will leave a truck looking shabby, regardless of the quality of the refinishing and restoration work. Brightmetal restoration and replating can be done at any time. Just be aware that rechroming services work at a rather slow pace; typically two months or more can elapse between the time you deliver the parts to the plater or ship them out and the time they are ready to be picked up or arrive back at your door.

Brightmetal Replating

The big question that most old-truck owners ask about chrome plating services is, How can I get the quality I want at an affordable price? The advice that follows may help you avoid the snags that can occur in the replating process.

Even though you can't see through the shiny chrome finish, the plating layer doesn't cover imperfections on the metal's surface. Any pits or damage to the metal will still be visible through the shiny new coating and indeed will be enhanced by the chrome finish. If you give the plater parts that are dented, rusted, or pitted, you will receive poor-quality plating unless the parts undergo a time-consuming—and expensive—restorative process of straightening, buffing, and filling the metal before plating. Platers will advise you—and their advice is to be heeded—to give them the best-quality parts possible; this will enable them to give you the best-quality plating at the most reasonable cost.

Let's say you own a Bonus Built Ford pickup with a plated grille (plating was available in 1948, the first year of the Bonus Built series). Most likely, the chrome grille bars have lost their shine and may even be spotted with orange rust that has bubbled up the plating in spots and eaten into the metal. If you're looking to have the grille bars replaced, you have three choices: have the bars taken through a restorative plating process (this expensive process is explained in more detail later in the "Brightmetal Plating Process" section) so that the finished product looks like new; have the plater just strip the bars of their old plating and put on new (this is a less expensive alternative, but you'll likely be disappointed in the results); or find a good set of painted bars and have the plater strip and replate them (the result should be like-new-quality plating).

Nothing brightens up an older truck like gleaming chrome and shiny brightmetal trim.

After the old plating is removed, parts to be replated are sent to the polishing room. Polishing is a manual operation requiring skilled labor. If the parts need only quick polishing to remove surface imperfections, the cost for this step will be low.

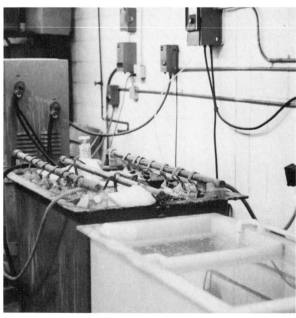

The chrome plating operation consists of several steps where the parts are immersed in electrolysis solutions for copper, nickel, and chrome coatings. After each plating "bath," the parts are thoroughly rinsed to make sure chemicals from one tank do not contaminate the next.

If you take the third option, make sure the paint is original and not a new coating hiding metalwork as well as pits. Another advantage of working with trucks as opposed to cars is that painted trim was often standard, with plating optional, and good painted parts replate nicely.

Brightmetal Plating Services

Your search for a plating service should be guided by the knowledge that two types of plating services are available. The first is commercial platers, which can be found through the Yellow Pages of telephone books for large cities. The second is restoration plating shops, which advertise in old-car and -truck hobby magazines.

Chances are, if you call a commercial plater about redoing the trim parts on your truck, the first question you'll be asked is whether or not any of these parts are diecast. Diecast is a synthetic metal that melts at low temperatures and therefore is ideal for molding into trim parts. It was used for years to make taillight brackets and housings, trim plates, hood ornaments, control knobs, and other decorative items on cars and trucks; more recently,

When parts emerge from the chrome plating tank, they are covered with a butter-colored film. The shiny chrome finish appears after the final rinse.

manufacturers have replaced diecast with plastic. Commercial platers are reluctant—most will refuse—to replate diecast, for some very good reasons.

First the metal is porous. This means that during the electrolysis process that is used to strip off the old plating and apply the new copper, nickel, and chromium layers, chemicals from the plating solution may seep into the metal. When this happens, the chemicals eventually work their way back out to the metal's surface, where they are trapped by the plating layer. Unable to escape, the chemicals will bubble up under the plating layer, giving the trim part the same appearance as a body part where rust in the metal has bubbled up the paint.

Second, when diecast corrodes, the rust doesn't spread across the surface, as it does with steel. Instead it eats into the metal. This means that corroded diecast can't simply be ground or buffed smooth. To restore corroded diecast, the rust actually has to be drilled out and the holes filled with brass or another noncorrosive metal; this process closely resembles filling a cavity in a tooth. A commercial plater does not have the time or in many cases the

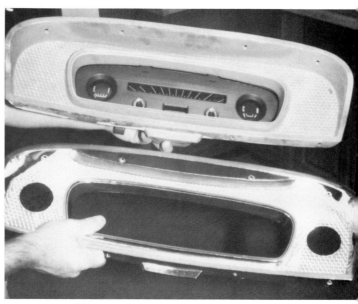

On sixties and later Ford trucks, some of the interior trim, such as the instrument panel shown here, has a chrome-plated look, but the base material is plastic. Note the bright finish of the replated part.

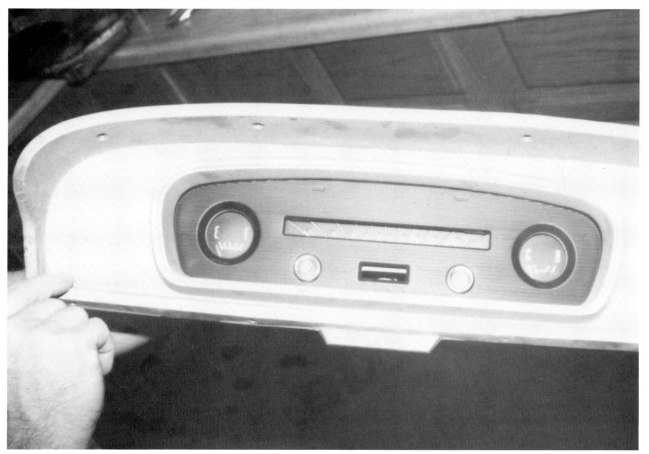

The guidelines for replating plastic parts are the same as for regular chrome plating. Select the best parts possible. In this photo, the plater is pointing out that if extra plating buildup was needed to coat a damaged area, the metal-turned texture on the face of the panel could be obscured.

203

skills to take diecast parts through this tedious and time-consuming restoration process—and the commercial plater is also savvy enough to know that anything less will not result in customer satisfaction. So in most cases, the commercial shop will simply say no to replating diecast trim parts.

Commercial platers will replate steel parts like bumpers and grille parts that are not diecast. For a truck that you plan to use as a "driver," commercial plating will give acceptable quality—if you bring in good parts. The best parts for commercial plating are bumpers or other trim that were painted originally and are not dented or badly rusted. The plater will remove the paint and any surface rust in a chemical dip, and then plate over fresh metal that had been protected by the paint finish.

As cautioned earlier, if you're looking for straight painted parts to have plated, don't choose items that have been repainted. There's no telling what condition the metal is in under the new paint. What you're looking for are bumpers and grille parts with a sound, original finish.

For replating diecast parts such as trim plates, hood ornaments, control knobs, and grilles on some thirties and early-forties trucks, you'll need to contact a restoration plating shop. Fortunately a truck does not have a lot of diecast metal, because replating these parts can get expensive fast.

Unless you deliver your parts to a restoration plating shop's display tent at a major swap meet, you will probably make contact by phone or mail. Typically you will be asked to send the parts for an estimate, then wait several weeks, possibly two to four months, for the plating to be completed.

Often you will have a choice of quality, with show plating being the most expensive but also giving the brightest shine and most perfect finish. In deciding on quality, consider more than just the cost. A better guide is how you intend to use your truck. If you plan to enter the truck in show competition, quality is important at every step of the restoration process. If you plan mainly to drive and enjoy your truck—and perhaps display it in local parades—then show-quality plating may not be desirable for two reasons. First, chances are, you won't wax and polish the truck with the fetish indulged on show vehicles, so the show-quality plating's brilliant shine will soon be hidden under a coating of dust. Second, if you have other parts—the bumper, for example— redone by a commercial plating service, the difference in quality between show chrome on the trim and commercial plating on the bumper will be noticeable. It's better that all brightwork be of a uniform grade.

Brightmetal Plating Process

To understand the differences in plating quality, let's look at what happens in the plating process. When the parts arrive at the shop, they are logged in and tagged. If any disassembly is needed, this is done next. Typically you'll be billed for labor if any disassembly is needed. This means you should do any disassembling before you send the parts for plating. For example, if you wanted to have the vent pane latches replated, you should pull them off the vent pane frame before sending the parts away.

As a first step in the plating process, the parts are immersed in a chemical bath to remove dirt, grease, paint, and other coatings. This done, the parts move through a series of baths where the old plating layers are stripped off by reverse electrolysis; chrome plating is removed in one tank, nickel in another, and so forth, until only bare metal remains. Next the parts go to a buffing room, where a polishing wheel removes any pits or other surface imperfections.

The polishing step is where the quality of the parts becomes important. Polishing is a manual operation requiring skilled labor. If parts need only quick polishing to smooth minor imperfections, the cost for this step will be low. But if the parts were

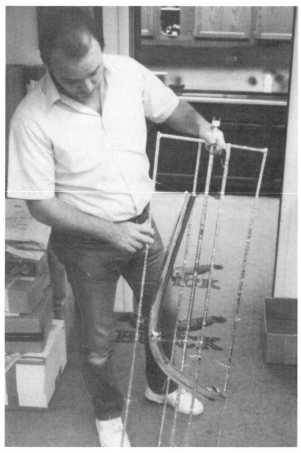

The plastic chrome plating process does not use chemicals. Instead parts to be plated are wired to a rack and placed inside a vacuum chamber, where they receive a very thin coating of aluminum. After plating—the process is actually called vacuum metalizing—the parts are given a coating of clear urethane to protect the shiny finish.

	BUFFING WHEEL & COMPOUND SELECTION CHART														
MATERIAL BEING BUFFED	STEEL, IRON, STAINLESS OR OTHER HARD MATERIALS			BRASS, COPPER ALUMINUM AND OTHER SOFT MATERIALS			CHROME PLATE NICKEL PLATE			SOLID AND PLATED GOLD AND SILVER			PLASTICS		
	STEP 1 ROUGH	STEP 2 INTERMED.	STEP 3 FINAL	STEP 1 ROUGH	STEP 2 INTERMED.	STEP 3 FINAL	STEP 1 ROUGH	STEP 2 INTERMED.	STEP 3 FINAL	STEP 1 ROUGH	STEP 2 INTERMED.	STEP 3 FINAL	STEP 1 ROUGH	STEP 2 INTERMED.	STEP 3 FINAL
COMPOUND EMERY	X														
STAINLESS		X						X							
TRIPOLI			X	X			NO ROUGH CUT RECOMMENDED			NO ROUGH CUT RECOMMENDED					
WHITE ROUGE			X			X		X						X	
JEWELERS ROUGE					GO TO STEP 3						X			GO TO STEP 3	
PLASTIC													X		X
WHEEL SISAL	X														
SPIRAL		X	X					X					X		
LOOSE SECT.			X			X			X						
FLANNEL											X				X

© The Easthill Group

Use this chart to determine the compound and wheel you'll need to buff or polish specific parts you're working on. The Eastwood Company

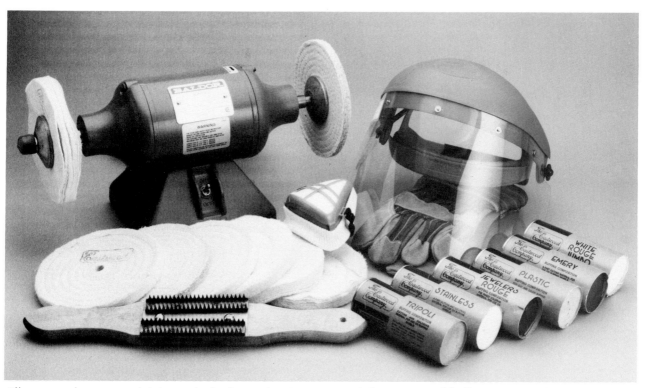

All you need to restore brightmetal trim is an electric motor, a set of buffing wheels, and an assortment of polishing grits and buffing compounds. The Eastwood Company

corroded and pitted, they'll take a lot longer to polish smooth. The added polishing time not only runs up the bill, but also results in some distortion to the part's shape, since the way the surface of the part is made smooth is by cutting away the metal.

In restoration plating, which is the costly alternative if good-quality parts can't be found, rather than remove pits and other surface scars by polishing, a soft copper plate layer is applied, then the parts are polished. In this process, the copper coating is used as a filler so that the part's shape isn't distorted. Commercial platers are interested in production and usually don't want to take the time for copper plating prior to polishing.

After polishing, the parts are placed in the copper plating tank. Copper is the base coating and the thickest layer in the so-called triple-plating process that underlays the chrome finish with layers of copper and nickel. Depending on what's desired for the final quality, the parts may stay in the copper tank for anywhere from 20 minutes to 4 hours. The penetrating bluish luster that seems to come from deep below the surface of show-quality chrome plating is the result of a thick copper base.

Between each plating step, the parts are rinsed thoroughly to keep chemicals in one tank from contaminating the next.

With restorative plating, the parts may be polished again to smooth any remaining surface flaws, and go back through the copper plating step yet another time.

The parts go next to the nickel plating tanks. Nickel plating ensures a good chrome bond and also helps give the chrome its deep, rich finish. The parts will spend about 20 minutes in the nickel plating tank.

Chrome plating is the final step and takes only a few minutes. When parts emerge from the chrome tank, they are covered with a butter-colored film. The shiny chrome finish appears after the final rinse. The chrome layer is the thinnest and least significant part of the plating process. As with the clear coat on a final paint finish, the brilliant shine and glossy smooth surface of a replated part are the result of care and thoroughness in the preceding steps.

Plastic Parts Replating

On sixties and later trucks, some of the interior trim—such as the dash, steering wheel hub, knobs, and air conditioning controls—have a chrome-plated look, but the base material is plastic. You might think that since plastic doesn't corrode, these parts would retain their shiny finish indefinitely. The enemy to these parts is sunlight. Over time, the chrome plating seems just to fade away. At first, it begins to dull. Eventually it simply disappears, and you're left with just the bare plastic.

Although the finish on these plastic interior parts looks like chrome, they're not really chrome-plated—at least not in the restoration process. The silvery finish is actually a thin layer of aluminum that is bonded onto the plastic in a process called vacuum metalizing. Several specialty companies offer this "plastic plating" service.

As in the normal chrome process, any surface imperfections in the plastic parts will show up in the shiny finish. For this reason, you should look for scratches, gouges, cracks, and breaks in the plastic. These imperfections can be repaired or filled by the plater, but this extra work adds to the plating expense. If the parts from your truck are in rather ratty condition, look for better parts. Beware of plastic with a white, powdery build-up on the surface. This build-up indicates that the plastic has lost all of its resins, or moisture. Plastic in this condition is considered dead. It is very brittle, which means that it can break very easily and should be handled with care.

The guidelines for vacuum metalizing of plastic parts are the same as those for regular chrome plating. Select the best parts possible. Dismantle assemblies—take the knobs off switches, remove the instruments from the dash unit, and so forth—so that you have only the plastic parts.

When you contact the plastic plating service for a price estimate, ask if the service applies a lacquer clear coat over the bright finish. The clear coat is necessary to protect the metalized finish from "evaporating," as occurred on the original part.

In most cases, you'll be shipping the parts to the plastic plater. Wrap the parts so that they are well cushioned against breakage. Typically it takes four to

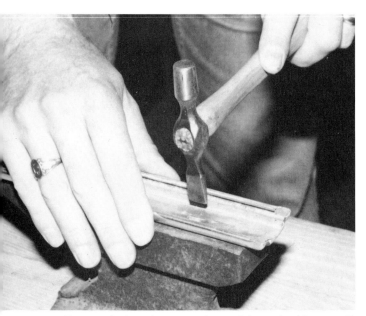

Damaged trim pieces can be straightened with a small-head trim hammer. Dents are worked out by tapping around the damaged area in a spiral motion, starting at the outer edge of the dent and working toward the center.

six weeks for a service to apply the bright finish and return the parts.

Brightmetal Polishing and Buffing

Much of the brightmetal trim on Ford trucks back to the early fifties is stainless steel or aluminum. This trim doesn't need to be replated, but has probably lost its original luster and may be dented or scratched as well. Although you can send it to a service that specializes in restoring automotive brightmetal, you can also do this work yourself at considerable savings. All you'll need is an electric motor, a set of buffing wheels, and an assortment of polishing grits and buffing compounds.

The expense to setting up a buffing station can be minimal if you have an extra electric motor on hand. Even if you decide to buy the motor as well as the buffing and polishing products, you'll still be money ahead if you decide to restore your truck's brightmetal trim yourself—and you will have the bonus of the satisfaction that comes from turning a battered part into a piece of gleaming metal that looks as if it just came from the factory. Polishing and buffing supplies are available from mail-order

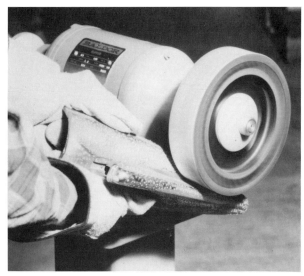

An expander wheel consists of an abrasive band fitted over a rubber wheel that is designed so that it can conform to contoured surfaces. The abrasive makes fast work of removing scratches from repairs to brightmetal parts, in preparation for the buffing steps that will restore the metal's shine. The Eastwood Company

When the surface feels flat, rub a fine file across the area you've been working. If the filing shows the surface to be essentially smooth, you can remove the deeper file scratches and hammer marks by sanding, either by hand or with an expander wheel.

restoration suppliers such as The Eastwood Company.

Before examining the process for restoring stainless and aluminum trim, let's understand what the terms *polishing* and *buffing* mean. When we think of polishing a finish, we think of applying wax and rubbing the finish to achieve a shine. But in the process of restoring trim, polishing isn't the step that brings out the shine; it's an earlier step that removes the deeper scratches and hammer marks left from straightening dents. This polishing step can be done by hand using progressively finer sandpapers. But if you're working with any quantity of trim, hand sanding would be slow and very time-consuming. A much faster approach is to polish out the scratches and mars using an expander wheel (a flexible wheel that bends to conform to the contours of the part) mounted on an electric motor.

Buffing is the series of steps that produces the shiny luster that is our goal on these brightmetal trim parts. The buffing process is done using a series of grits applied with first coarser and then softer wheels mounted on the electric motor.

Brightmetal Polishing Process

The steps in restoring a damaged piece of brightmetal trim, such as a side trim spear on a mid-sixties Ford pickup, begin with the removal of the trim piece.

The next step is to straighten any kinks or dents. This is done using a flathead bolt, a hardwood dowel, or a small-head trim hammer. The way to work out dents is to tap around the concave area in a spiral motion, starting at the outer edge of the dent and working in toward the center. When the surface feels and looks flat, rub a fine file across the area you've been working. Don't cut deeply into the metal. The purpose of doing this is to check for remaining high and low spots. You'll see the high spots by where the file cuts into the metal.

If the surface is still somewhat rough, you'll need to do more straightening. But if the file shows the surface to be essentially smooth, you can proceed to remove the deeper file scratches and hammer marks by sanding, either by hand or with an expander wheel. You'll start with 220-grit sandpaper. If you are hand sanding, use a wooden or rubber block. With stainless trim, you'll follow the 220-grit with 360-grit sandpaper, but with aluminum, you can proceed to the buffing process.

If you are doing the sanding, or polishing, process with an expander wheel, each time you change sanding belts, you should work the part at a right angle to the previous passes. Be sure to wear gloves to protect your hands from heat buildup on the parts and possible contact with the sanding wheel, and always wear a face shield to protect your eyes and face from flying debris.

Brightmetal Buffing Process

Polishing will bring the trim to a smooth but not shiny finish. The buffing steps restore the shine. With hard metals like stainless, the buffing process begins by installing a sisal wheel (designed for fast, aggressive cutting) on the arbor of the electric motor and coating the wheel with a coarse emery grit. The part should be buffed in small sections until all areas have been worked. When using a rough wheel and

Buffing is done with a series of progressively milder grits and softer wheels. The coarser grits remove surface oxidation, and the milder grits restore the brightmetal's shine.

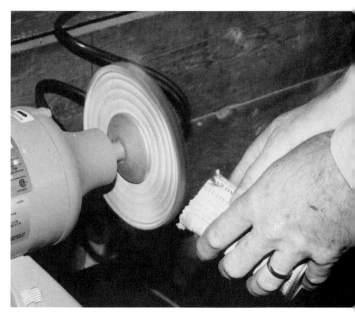

The buffing action is accomplished by compounds of various grits, which are applied to the wheel periodically during the buffing process.

coarse grit, be alert to rapid cutting that could distort the part's contours or cut holes in the metal. After the part has been coarse buffed, a softer spiral wheel is installed and the piece is buffed with a milder grit. This intermediate buffing is followed by a finish buffing using a soft loose section wheel and a coloring compound to bring the metal to very high luster.

The preferred buffing setup is a double-arbor motor with a coarse wheel installed on one end of the shaft and a softer wheel on the other. This way, it is not necessary to change wheels between buffing steps.

Beginning buffers typically apply too much grit to the wheel and forget to clean the wheel before going to a milder grit. Not only is putting too much compound on the wheel wasteful, but the extra compound also clogs the wheel and hinders the buffing action. Black streaks on the work piece are signs of too much compound. These streaks can be cleaned off with a soft cloth soaked in solvent. The easiest way to make sure that none of the harsher compound remains on the wheel as you move to a milder grit is to use a different wheel for each buffing step. Otherwise you can remove leftover compound from the buffing wheel with a tool called a rake.

One other important caution: Never take your eyes off the part while buffing.

Plastic and Aluminum Polishing and Buffing

Plastic parts like taillight and parking light lenses can also be restored by buffing. Typically these plastic lenses have a dull look caused by tiny scratches. Using a soft spiral wheel and special plastic buffing compound, followed by a finish buffing with a flannel wheel, the scratches can be removed, giving the plastic a shiny, new look. When buffing plastic, be careful not to remove part numbers or other detail features.

Polishing and buffing can also be used to produce a chromelike finish on aluminum parts like alternator housings and intake manifolds. These parts did not come from the factory with a shiny finish, so buffing them is not to be done as part of restoration. Shiny engine compartments, however, appeal to street rodders and others. Handheld grinders can be used to polish small recesses on these parts, that can't be reached by the expander wheel.

Although trucks don't have a lot of chrome and brightmetal, replating the chrome and buffing the aluminum and stainless trim pieces will greatly improve the appearance of your truck. If yours is a standard model truck with mostly painted trim, you may also decide to add some of the brightwork found on the deluxe models. This would not be deviating from the original, as the chrome trim, when available, was an extra-cost option.

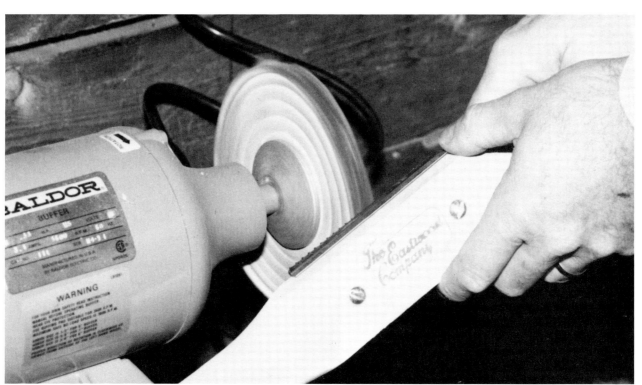

Beginning buffers often apply too much compound and forget to clean the wheel before going to a milder grit. Not only is putting too much compound on the wheel wasteful, but the extra compound also clogs the wheel and reduces the buffing action. Excess compound is removed by a tool called a rake. This tool is also used to clean the wheel between buffing steps.

Chapter 17

Storage and Care

Once we've finished with our truck, it's likely to spend more time in storage than it does traveling down the highway. Walking past our truck in the garage, to get into our everyday vehicle, may give us a good feeling—a sense that the old pickup is enjoying a well-earned rest that will help it last forever—but in reality, storage can do more damage to a collector vehicle than does regular driving.

A lot of the destruction occurring during storage traces to the conditions in which the truck is parked. But the very failure to drive the truck can be harmful in itself. If the truck is parked in a barn or shed that is also home to rodents, all kinds of damage can result—from chewed wiring to nests in the upholstery, if the mice or squirrels can get into the cab. If the truck sits outside or is parked on a dirt or concrete floor, moisture will collect on the underside, causing rust to develop faster than you'd

Rule number one for any storage period is not to park your truck outside. Moisture from the ground will rust the underside while sunlight fades the paint and hardens the rubber.

like to think. In storage, the battery will discharge unless it is recharged periodically. If the battery sits for very long in a fully discharged condition, you're likely to need a new one when you go to start the truck.

This chapter gives some practical guidelines that will help your truck come out of storage in as good a condition as it went in.

Long-term and Short-term Storage
Storage can be thought of in two categories: long-term and short-term. Long-term can mean the winter months (November through March, let's say) or an even longer period—perhaps years. Most show trucks exist in a state of long-term storage, only being started and driven out of their enclosed trailer onto the show area, then back into the trailer at the end of the show. Short-term storage is the time your truck spends sitting in the garage between occasions when you take it out on the road. If your purpose for owning an older truck is to drive it, your truck may only sit a few days at a time. For short-term storage, you're mainly concerned with keeping the truck dry. Long-term storage has a longer list of conditions if you're going to keep the truck from deteriorating while it is sitting.

Rule number one for any storage period is not to park the vehicle outside. Moisture from the ground rusts the underside while sunlight fades the paint and hardens the rubber. Condensation forms on the inside of the doors and under the hood, causing not only rust, but also corroded wiring connections and stuck engine valves. Putting a vehicle out to pasture is a quick way to blend it in with the landscape.

Indoor storage is not a guaranteed cure-all. A roof over a dirt floor can cause more moisture damage than if the truck is left outside—especially if the roof is metal. Ideal storage is warm, dry, and dark, with very little fluctuation in temperature. Unfortunately not many of us have a storage facility that meets these ideal requirements. Happily an inexpensive device can create the ideal storage conditions regardless of what the building where we park our truck is like—even if it's a tin-roofed, dirt-floored shed.

Omnibag Storage

The simplest and least expensive way to put your truck into long-term storage under ideal conditions is to purchase a product called an Omnibag. As the name implies, this is an all-encompassing bag—a huge gray plastic bag that's big enough to roll your truck into. When the truck is inside the bag, you'll toss in several packets of desiccant (moisture remover), then roll up the opening of the bag and seal it shut. The desiccant will absorb whatever moisture is trapped in the bag, and the sealed plastic will keep any more moisture from entering.

Once inside the bag, the truck will stay as dry as if you had parked it in a desert, for the duration of its storage. Moisture can't creep up from underneath because the truck is sealed from the floor by the bag. Changes in temperature or outside humidity can't soak the truck in beads of sweat because the desiccant has removed the moisture from the bag. Sunlight can't harden the rubber or dull the paint because the bag keeps the truck in the dark. Rodents won't chew the truck's wiring because mice and squirrels don't go after smells they don't like and they apparently aren't attracted to the aroma of the bag. Moisture won't enter the engine because, as already stated, the bag seals your truck in a no-humidity environment.

The only requirement for using an Omnibag is that the storage be inside. The bag is not recommended for out-of-doors use because wind will whip the plastic, causing tears that will let in moisture, which will then be trapped inside the bag. In this condition, what was formerly the best storage situation becomes the worst. With indoor storage, if you accidentally make a small tear in the bag, it can be mended with duct tape and the ideal storage conditions will be restored. With care, the Omnibag can be reused year after year. The Omnibag's manufacturer, Pine Ridge Enterprises, is listed in the Appendix.

Long-term Storage Guidelines

Before placing your truck in an Omnibag, or any long-term storage setting, a few preparation steps should be taken. It's advisable to pump up the tires to slightly over road pressure, and the truck should be washed underneath and on top and allowed to dry thoroughly.

You'll sometimes hear the advice to place the truck on jack stands and remove the wheels and store them in a dark, dry place. This really isn't necessary—and isn't possible if you are storing your truck in an Omnibag—because even if your truck has bias-ply tires (the older-style tires used before radials became popular), which will develop flat spots in storage, they will round out again in a few miles of driving.

If you wax the vehicle's finish, then when the truck comes out of storage, it will be shiny and ready to go or show. If you put your truck in an Omnibag, it won't need to be washed. In other storage settings, enough dust will collect on the truck that you'll want to give it a wash-down before the first outing.

To protect the chrome, you can apply a wax coating and then let it dry and leave the white powdery film. The wax will protect the chrome from oxidation. This step is also not necessary if the truck is stored in an Omnibag.

The interior should be cleaned as well. This means vacuuming the floor mat, and washing the seat and painted interior surfaces with a bucket of water into which you have squeezed a couple of squirts of dishwashing detergent. After washing, rinse off with clean water.

You can protect the engine against rusting during storage by starting it up and *slowly* pouring about 1/2 cup of clean engine oil down the carburetor throat. You will need to advance the carburetor above idle to keep the engine from stalling, and once the oil is in the cylinders (you'll

The simplest and least expensive way to put your truck into long-term storage under ideal conditions is to purchase a product called an Omnibag. Basically this is just a gray plastic bag large enough to hold a car or truck. Here we've unrolled the Omni, which is the first step to "bagging" the Model A in the background.

see smoke rolling out the exhaust), you can shut off the ignition. The oil will form a coating on the cylinder walls to prevent rust and guard against ring seizure.

It's important to add gasoline stabilizer to the fuel tank. This will help prevent varnish from settling out of the gasoline and clogging the fuel lines and filters.

The radiator coolant should be checked to make sure it contains enough antifreeze to protect against freezing. Draining the coolant and storing the truck with a dry cooling system is not advised. Today's antifreeze contains rust inhibitor, which wards off corrosion inside the radiator and engine coolant passages.

If the truck is not stored in an Omnibag, it's a good idea to tie bread bags over the exhaust outlet and carburetor to seal out moisture that will otherwise enter through these openings. Also if an Omnibag is not used, the truck should be covered to protect the finish from dust and to keep out sunlight.

The last step before rolling the truck into an Omnibag or covering it for its long-term slumber is to take out the battery. Although a battery in a vehicle that is driven regularly requires very little care, a battery in storage will deteriorate rapidly. Battery care has its own set of conditions, which are listed in the "Battery Care" section later in this chapter.

Short-term Storage Guidelines

Short-term storage should also be inside, if possible, with the vehicle covered. It's not advisable to cover vehicles stored outside, for the same reasons that the Omnibag manufacturer does not recommend that its product be used for outdoor storage. A cloth cover may not tear, but it will whip in the wind, and as it rubs against dust that collects on the truck's finish, the paint is going to be scratched and abraded.

For short-term storage, it is advisable to install a battery disconnect switch (available from Bathurst, Inc.; see the Appendix for contact information) to prevent the battery from discharging owing to a possible short in the electrical system. If the short-term storage lasts for several weeks, it's a good idea

With the Model A nearly enclosed by the Omnibag, we will toss in several packages of desiccant. Then we'll roll up the opening and seal the bag shut. The desiccant will absorb whatever moisture is trapped in the bag, creating a desertlike storage condition.

to trickle charge the battery every few days during the storage interval.

Removal from Storage

Even though a truck stored in an Omnibag will come out of storage in virtually the same condition that it went in (the chrome will still be bright and the finish won't be dusty), a few minutes spent checking the vehicle over can avoid problems when you're on the road. Gaskets and seals can shrink during storage, especially in a low-humidity environment, so always check fluid levels for the engine, transmission, brake reservoir, and coolant before starting and driving the truck. You'll need to replace the battery, and you'll find that pouring a small amount of gasoline down the carburetor intake and maybe spraying a couple of shots of carburetor cleaner into the carburetor as the engine first cranks over will save grinding on the starter while the fuel pump refills the carburetor float bowl. Once the truck is running, let it idle a few minutes to lubricate the engine, before moving the vehicle from its resting place. If you've stored the truck in an Omnibag, you can fold up the bag and put it away for the next storage season.

Trucks with bias-ply tires will ride a little bumpy for the first few miles until the tires warm up and round themselves out again. After the first outing, check the fluid levels again and top up where necessary. Leaks caused by dried-out gaskets will probably stop as the truck is driven, but if the fluids are down, it's a good idea to keep monitoring them until the fluid levels stabilize.

Truck Care

A collector truck that is driven a few hundred miles a year may seem to need very infrequent maintenance. But this is not really the case. Corrosive chemicals that collect in the crankcase as the engine oil breaks down are actually more damaging in a vehicle that is driven infrequently than in one that goes out on the road every day. So change your truck's oil and do a complete lubrication at least once a year—regardless of how few miles it is driven. The best time to do this is right before putting the truck in long-term storage. Trucks that see more than occasional use should be given a lube and oil change every 2,500 to 3,000 miles.

If you look at the lube chart in your truck's owners manual, you'll probably see that your pickup has several dozen grease fittings and that these are not just on the steering linkage and front suspension, where you would expect grease fittings to be, but also on the spring shackles, universal joints, clutch and brake pedal shafts, and other places you probably wouldn't think to look. To make sure all of these fittings are greased, it's a good idea to lube the truck yourself instead of taking it to a quick-change oil-lube place. Doing a complete lube will pay off by a smoother ride and drive. While servicing the

Whenever a vehicle is in storage, the battery needs to be kept charged. The easiest way to do this is with a solar battery charger. The solar panel can be placed on a window ledge, or near any light source, and left connected to the battery. The solar charger's small, 1amp output will not overcharge the battery.

engine and chassis, remember to squirt a couple of drops of oil on the manifold heat valve and generator bushing cup; note the instructions in chapter 13 about putting a few drops of oil in the speedometer head on Bonus Built trucks; and check the fluid levels of the steering box, transmission, rear end, and brake master cylinder.

The annual maintenance check is also the time to wash the oil filler cap in solvent, on engines built before positive crankcase ventilation (PCV) systems, and add a few drops of oil to the wire mesh. The filler cap and screen ventilate the crankcase.

The carburetor air cleaner—whether oil bath, wire mesh, or paper filter type—should also be serviced at this time. An oil bath filter is serviced by dumping out the old oil (pour it into a plastic milk jug or other suitable container and take it to a recycling center; don't pour it onto the ground), cleaning the oil reservoir and filter in solvent, and then refilling the filter bath with fresh oil. A wire mesh cleaner is washed with solvent and dried; squirt a few drops of oil into the mesh before replacing this type of air cleaner on the carburetor. On engines using a paper air filter, you'll probably want to remove and wash the filter housing. The filtering sponge at the PCV valve hose inlet should also be washed in solvent or replaced. The new paper filter cartridge is installed after the air cleaner housing is replaced on the carburetor. On newer engines, remember to check the PCV valve and replace it if it's clogged.

Like engine oil, antifreeze solutions contain

additives that break down in time. This means that if the cooling system is not flushed periodically (at least biannually) and refilled with fresh antifreeze, internal corrosion can result.

Many collectors are also unaware that manufacturers recommend flushing and refilling the hydraulic brake system on an annual basis. The reason for this is that polyglycol brake fluid soaks up moisture like a sponge and the master cylinders on older trucks are vented to prevent a vacuum from forming in the system, so moisture is easily absorbed into the brake fluid. The best way to avoid this moisture problem—and to save the time flushing and refilling the brake system—is to use DOT 5 silicone fluid instead of DOT 3 polyglycol hydraulic brake fluid.

You can wrap up the truck care session by putting a few drops of oil on the hood latch mechanism; hood hinges; rotary door latches; door hinges; seat track, on trucks with adjustable seats; and wiper linkage. The door latch striker should be lubricated with special stick wax.

Cleanup Tips

You may look at this heading and say, "Hold it, I don't need to be told how to wash my truck." That's probably true, but still, you may discover a few techniques and "tricks" that you can apply in caring for a collector vehicle but haven't used in cleaning your daily drivers.

When washing your vintage truck, the low-pressure water from the yard hose is recommended over a high-pressure spray at the car wash. The high-pressure spray can force dirt and debris into seams and crevices—making a breeding spot for rust. If you've restored your truck, you'll also want to wash it underneath as well as on top.

For cleaning chrome and removing water spots from the finish, removing road tar, and handling any number of other cleanup situations, you'll find the line of car care products from One Grand Car Care Products (see Appendix for address) to be far superior to those you might otherwise pick up at the discount mart.

A simple trick makes windshield and cab window glass look sparkling clean: just add a dash of kerosene to a bucket of clean water, then wipe the mixture on the glass, and buff dry with a wad of newspaper. The combination of the kerosene film and newsprint will give the glass a mirror sheen.

Rubber, including the tires, can be restored to its factory-bright black coloring with a coating of Armor All—which also serves to protect the rubber against cracking.

On the inside of the cab, a damp cloth or sponge works well to clean vinyl seat coverings and to remove dust from door panels and the headliner. The rubber floor mat can be vacuumed, wiped down with a damp cloth, and renewed to its factory freshness with Armor All. Other interior rubber—the gearshift boot on trucks with a floor shift, the pedal pads, and other items—should also be given the rubber renewal treatment.

Under the hood, you can keep the engine compartment clean by spending a few minutes wiping off road dust and whatever oil film has developed, each time you wash the truck. If the engine is too grimy to be cleaned with a light wipe-down, Gunk degreaser will bring back that factory-fresh look. Gunk works best in warm temperatures, so running the engine to warm up the surfaces will encourage its degreasing action. After spraying Gunk on the grease-coated areas, let it work for 10 to 15 minutes, then wash off the grease film with the lawn hose. Be careful not to spray water on the distributor or other electrical parts. Used according to the directions on the can, Gunk will not harm paint.

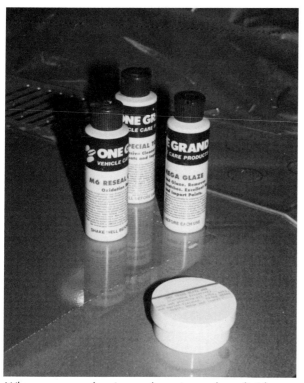

When you are cleaning and waxing a show finish, you will want to use products that are nonabrasive. The cleaners and waxes from One Grand Car Care Products maintain the finish chemically and without abrasives. This company markets a system of products for all cleaning and shining needs.

Routine Maintenance

If your truck's engine was built before unleaded gasoline became popular in the seventies, and has not been rebuilt with hardened valves and seats, it's wise to add a valve lubrication additive at gasoline fill-ups. An engine that was designed for gas containing tetraethyl lead can develop very rapid valve seat wear when fueled with unleaded gasoline.

Check the wheel bearings for looseness and lubrication, clean and regap the spark plugs, file the points, and pack an extra set of points, condenser, and fan belt and a basic tool set before taking a trip of any distance. Just because your truck spends most of its time sitting in a garage doesn't mean that it's in road-ready condition. Spending a couple of hours giving your truck a thorough check-over before heading out of town is good insurance that you won't be stranded on the highway with fried ignition points or a dry wheel bearing.

Tips for Showing Your Truck

In preparation for show competition, cleaning and waxing needs to be done with extreme thoroughness—to the extent that you will even wipe wax paste out of cracks with a toothbrush and make sure the chassis is as spotless as the exterior. The secret of success in show competition is paying close attention to the many small details that make your truck authentic for its period. This means that if your truck was built before the early fifties, the valve stem caps should be metal, not plastic, and the metal base on the spark plugs should be painted black.

Paying attention to detail includes arranging an attractive display. If you have fitted your truck with unusual accessories, the display should include dealer literature showing these accessories. Then, if the judges who are evaluating your truck question the authenticity of these accessories, you will have the documentation needed to keep your truck from losing points. If you mount the accessory literature on a display stand that you can place near your truck, the information can also seen by those viewing your vehicle. Serious competitors will even rotate the wheels so that the Ford script on all hubcaps aligns horizontally.

A tip for those entering prestigious AACA competition: You are entitled to request a judging evaluation after each competition round. You will not be given a copy of the actual judging sheet, but you will be told in a general way where you lost points. By correcting these errors before the next competition, you will greatly improve your chances for success.

Battery Care

Batteries in collector vehicles that are stored more than they are driven typically have a very short life. This isn't necessary. You'll get many years of life and service out of your truck's battery if you follow these three simple steps:

1. Keep the case clean and dry.
2. Keep the level of the electrolyte (the fluid in the cells) above the plates.
3. Keep the battery fully charged.

Dirt and grease on the case create an electrical circuit that can cause the battery to discharge in a relatively short time. When you place the truck in storage, remove the battery and wash the case with a cloth soaked in ammonia or a baking soda and water solution. When washing the case, be very careful that none of the solution gets into the cells, as it will neutralize the electrolyte. Before storing the battery, be sure the case is wiped dry. Don't place the battery in a location, such as on a cement floor, where it will pick up condensation.

Some electrolyte is lost each time the battery is charged. Make sure the electrolyte never falls below the top of the plates. If the plates are exposed to air, permanent damage to the battery will result. Distilled water should be used to fill the battery. On most batteries, the correct electrolyte level is indicated by a ring, 1/4in to 1/2in above the plates, under the cell cap.

Many old-truck owners replace batteries needlessly because they fail to keep the batteries charged. Batteries discharge when they're not used, so they need to be recharged at fairly frequent intervals. This is done with a battery charger. A slow charging rate of 4amp to 6amp is best for the battery. It takes more than a few hours to bring a discharged battery up to full charge. Using an inexpensive home-shop charger, bringing a battery up to full charge may take a day or more.

The simplest and most accurate way to tell when a battery is fully charged is to take a reading with a hydrometer (a rubber device that looks like a thermometer and is available for a few dollars from an auto supply store). Most battery chargers also show when the battery is reaching full charge, by cutting back the amp output.

Be sure the battery can "breathe" when

The secret to success in show competition is paying close attention to every small detail. The owner of this Ford F-100 telephone truck has fitted out his vehicle with telephone installer's or repairer's equipment, authentic to the truck's era.

recharging. This means that the hydrogen gas created during the recharging process can escape from the cells. The vent caps contain small holes for this purpose. But sometimes these holes become plugged. If this happens, pressure can build up inside the battery and an explosion can result. To make sure the gas can escape, but also to avoid any risk that the hydrogen gas can collect and be ignited, remove the cell caps and place a damp cloth over the top of the cells. The cloth will absorb the hydrogen gas and allow it to combine harmlessly with the water molecules. After the battery has been recharged, remove the damp cloth, dry the top of the battery case, and replace the cell caps.

Before replacing the battery in the truck, clean the posts and cable clamps. An easy way to prevent corrosion at these contact points is to give them a coating of Plastidip. This rubberizing product is available at discount marts. The Plastidip coating will seal out moisture and corrosive vapors that are given off during the discharge and the recharging process, and preserve a good contact between the cable clamps and posts. You can also give the hold-down frame and bolts a coating of Plastidip to prevent these parts from corroding.

Maintenance Tips for Batteries in Stored Vehicles

•Charge the battery every two weeks.
•Store the battery in as cool a place as possible—to help minimize self-discharge.
•Keep the case clean and dry.
•Make sure the electrolyte stays above the top of the plates.
•Install a shutoff switch or remove the battery cables while the vehicle is in storage, to prevent possible shorts in the electrical system from discharging the battery or starting a fire.

When you are washing your vintage truck, the low-pressure water from a yard hose is recommended over a high-pressure spray at the car wash. The high-pressure spray can force dirt and debris into seams and crevices, making a breeding spot for rust. Noel Glucksman

Appendices

If you have recently purchased a vintage Ford pickup, you're likely to have questions about sources or parts, or may want to make contact with others who own similar trucks. The club, parts, and suppliers listings that follow will help with parts and supplies information as well as link you with others who share your interest. You will also find a listing of specialists in horn repair, lock rekeying and repair, instrument refinishing, and other services of this nature. Literature venders and further reading listings give you sources of historical and other information to help you restore your Ford pickup.

Clubs, Publications, and Events

Early Ford V-8 Club of America
P.O. Box 2122
San Leandro, CA 94577

This international club supports the preservation of all early V-8 Fords, including pickups.

F-100 Supernationals
c/o Pat Ford Promotions
1920 Council Ave.
Statesville, NC 28677

This is the world's largest gathering of Ford pickups, held annually at Pigeon Forge, Tennessee. It is a family-oriented gathering of trucks and parts venders, with technical workshops, show competition, parades, and related events.

'49-50-51 Ford Owners Newsletter
1733 S. Willow Dr.
Midwest City, OK 73130

Although not a club, this publication conducts many club functions, including an annual national meet. Benefits to Ford truck owners include a column on Ford trucks and the occasional publication of Ford truck service bulletins.

Light Commercial Vehicle Association (LCVA)
P.O. Box 838
Yellow Springs, OH 45387

The LCVA is national in scope, with members in Canada, western Europe, the Near East, and Australia. This group's charter embraces all makes of light trucks, but the club gives particular attention to Ford pickups—with a regular Ford column in its bimonthly publication and frequent Ford technical and feature articles.

Model A Ford Club of America
250 S. Cypress
LaHabra, CA 90631

Model A Restorers Club
24822 Michigan Ave.
Dearborn, MI 48121

These international clubs support the preservation of Model A Fords of all types, including pickups and larger trucks.

Model T Ford Club of America
41 Reeland Ave.
Warwick, RI 02886

Model T Ford Club International
P.O. Box 438315
Chicago, IL 60643

These international clubs support the preservation of Model T Fords of all types, including pickups and larger trucks. Both clubs offer magazine-quality bimonthly publications, hold national meets, and have regional chapter**s**.

Parts

Bob Drake Reproductions
1819 N.W. Washington Blvd.
Grants Pass, OR 97526

Reproduction parts for 1932–66 Ford pickups

Coachworks of Yesteryear
P.O. Box 651
Oakland, FL 34760

Complete reproduction 1936 Ford pickups and 1936 fiberglass replacement fenders, cabs, doors, and other body parts

Dennis Carpenter Ford Reproductions
P.O. Box 26398
Charlotte, NC 28221-6398
    Parts of all kinds for Ford pickups, 1932–66

Fairlane Company
210 E. Walker St.
St. Johns, MI 48879
    Quality fiberglass replacement fenders, hoods, and running boards, and custom bubble skirts for 1948–52 Ford F-1 and 1953–56 and 1961–66 Ford F-100 trucks

Gaslight Auto Parts
P.O. Box 291
Urbana, IL 43078
    Reproduction steel fenders and other Ford parts, Model T era through 1948

LeBaron Bonney Company
6 Chestnut St.
Amesbury, MA 01913
    Upholstery kits for Ford pickups, Model A era through 1952

Mack Products
100 Fulton Ave.
P.O. Box 278
Moberly, MO 65270
    Parts and complete pickup boxes

Obsolete Ford Parts
66015 Shields Blvd.
Oklahoma City, OK 73149
    Parts for Ford pickups, Model T era through 1972

SoCal Pickup
6321 Manchester Blvd.
Buena Park, CA 90621
Parts for 1953–56 Ford F-100 pickups

Performance Parts
Daniel R. Price
7320 Sunbury Rd., Rt. 2
Westerville, OH 43081
    Thomas finned aluminum cylinder heads and other speed equipment and parts for Ford Model As and Bs

Fifth Avenue Antique Auto Parts
502 Arthur Ave.
Clay Center, KS 67432
    Speed equipment for Ford and Mercury flathead V-8 engines

Patrick's
Box 648
Casa Grande, AZ 85222
    Speed equipment, including new Mallory dual-point distributors

Suppliers
Auto Mat Company
225A Park Ave.
Hicksville, NY 11801
    Carpet sets for 1953–70 Ford pickups and the Bronco; pickup bed mats and liners

Automotive Surplus
4920 S. Monroe St.
Fort Wayne, IN 46806
    Large inventory of new and restored bumpers

Bathurst
Box 27
Tyrone, PA 16686
    Battery disconnect switch

T. N. Cowan Enterprises
P.O. Box 900
Alvarado, TX 76009
    Wet/Dry paint stripper, Zintex rust neutralizer, and other chemical restoration products

Custom Autosound
808 W. Vermont Ave.
Anaheim, CA 92805
    AM-FM radio with stereo cassette conversion for older Ford trucks

The Eastwood Company
580 Lancaster Ave.
Malvern, PA 19355
    Extensive line of specialty tools and literature for automobile and light-truck restorers

Fifth Avenue Antique Auto Parts
502 Arthur Ave.
Clay Center, KS 67432
    Six-volt alternators—reliable, higher-output power sources for your truck's electrical system

Fort Wayne Clutch
2424 Goshen Rd.
Fort Wayne, IN 46808
    High-performance, antique, or obsolete clutches; rebuilt or made-to-order drive shafts

Gear Vendors
1035 Pioneer Way
El Cajon, CA 92020
    Modern underdrive-overdrive transmissions for late-model Ford trucks

GLASCO
85 James St.
East Hartford, CT 06108
    Replacement flat windshield and door glass

Bruce Horkey Cabinetry
R.R. 4, Box 188
Windom, MN 56101
    Stock or custom pickup bed floors and complete boxes

Mar-K Specialized Manufacturing Company
8022 N. Wilshire Ct.
Oklahoma City, OK 73132
    Reproduction pickup bed parts

Metro Moulded Parts
11610 Jay St.
P.O. Box 33130
Minneapolis, MN 55433
    Large inventory of reproduction rubber parts

MikeCo
c/o Michael Anthony Varosky
1901 Colonia Pl., #B
Camarillo, CA 93010
    Pre-1975 auto and truck lens replacement

NAPA auto parts stores nationwide
    Very large inventory of parts for modern and vintage vehicles, available overnight from regional warehouses

One Grand Car Care Products
13820 Saticoy
Van Nuys, CA 91402
    Complete line of cleaners, polishes, and waxes to maintain and renew finishes, to clean and preserve chrome, and to meet other truck care needs

Overdrives
17518 Euler Road
Bowling Green, OH 43402
    Manual control for Borg-Warner overdrive transmission; installation of overdrive transmission in pre-1942 Ford pickup torque tube drive

Pine Ridge Enterprises
13165 Center Rd.
Bath, MI 48808
    Omnibag—a dry storage device that protects the entire truck during long-term storage

Restoration Specialties
P.O. Box 328, R.D. 2
Windber, PA 15963
    Weather strip, fasteners, hardware

Rhode Island Wiring Service
P.O. Box 3737
Peace Dale, RI 02883
    Wiring harnesses

Solar Electric Engineering
175 Cascade Ct.
Rohnert Park, CA 94928
    Maintainer—a solar battery charger

Lynn Steele
1601 Hwy. 150 E.
Denver, NC 28037
    Large inventory of reproduction rubber parts

TiP Sandblasting Equipment
7075 Route 446
P.O. Box 649
Canfield, OH 44406
    Sandblasting equipment of all types and sizes

Y n Z's Yesterday Parts
333 E. Stuart Ave., Unit A
Redlands, CA 92374
    High-quality replacement wiring harnesses

Services
American Plastic Chrome
1398 Marann
Westland, MI 48185
    Chrome interior plastic parts

Bill's Speedometer Shop
3353 Tawny Leaf
Sidney, OH 45365
    Provides quality gauge repair

Eugene Gardner
10510 Rico Tatum Rd.
Palmetto, GA 30268
    Offers a large selection of good- and better-quality license plates, singles or pairs

Gas Tank Renu—USA
12727 Greenfield
Detroit, MI 48227
    Repairs and restores gas tanks to better-than-new condition; has franchises in the East and Midwest; is expanding nationwide

The Horn Shop
7129 Rome-Oriskany Rd.
Rome, NY 13440
    Repairs horns for most models

The Key Shop
144 Crescent Dr.
Akron, OH 44301
    Repairs and rekeys all locks

Photo Card Specialists
1726 Westgate Rd.
Eau Claire, WI 54703
Will make very high-quality color business cards based on a photo of your truck, as well as color photo post cards, and calendars

Literature Venders
Classic Motorbooks
P.O. Box 1
Osceola, WI 54020
Huge selection of automotive and truck books and manuals, including dozens of titles covering Ford trucks

Dragich Auto Literature
1660 93rd Lane NE
Minneapolis, MN 55434
Discount books

Walter Miller
6710 Brooklawn Pkwy.
Syracuse, NY 13211
Huge assortment of original 1910–80 sales literature, and repair and owners manuals for Ford trucks and other vehicles

Further Reading
Ford Pickup Repair: Pickups of the '70s
Contains reprints of *Hot Rod* magazine articles on performance tuning Ford F-series two-wheel-drive trucks of seventies vintage. Available from Classic Motorbooks.

Ford Pickups, 1932–52
by Mack Hils
Gives excellent photo coverage of Ford pickups of this period. Available from Mack Products.

Ford Pickups, 1957–67: How to Identify, Select, and Restore Ford Collector Light Trucks, Panels, and Rancheros
by Paul G. McLaughlin
Provides excellent historical coverage of Ford trucks of this period. Contains photos, ad reprints, and helpful lists of specifications. Available from Classic Motorbooks.

Ford Pickup Trucks: Development History and Restoration Guide, 1948–56
by Paul G. McLaughlin
Supplies excellent historical coverage of Ford trucks of this period. Contains reprints of original ads and helpful lists of specifications. Available from Classic Motorbooks.

Heavyweight Book of American Light Duty Trucks, 1939–1966
by Tom Brownell and Don Bunn
Offers comprehensive coverage of all American light trucks in this period; contains restoration information on renewing hydraulic brakes, replacing wiring, and rebuilding a straight axle front end. Available from Classic Motorbooks.

Hollander interchange manuals
Includes parts interchange information for domestic cars and light trucks, from the twenties to the present. Available from Dragich Auto Literature.

How to Restore Your Collector Car
by Tom Brownell
Presents step-by-step procedures for restoring a car or light truck. Includes topics such as stripping, derusting, metal repair, priming, painting, security measures to prevent theft, preparation for show competition, and more. Available from Classic Motorbooks.

Illustrated Ford Pickup Buyer's Guide
by Paul G. McLaughlin
Gives a year-by-year look at Ford light trucks, with value and appreciation ratings. Available from Classic Motorbooks.

Standard Catalog of American Light Duty Trucks, 1896–1986
Contains specifications for Ford and other light-duty trucks. Available from Krause Publications.

Color Options and Applications
The lists that follow give colors available for Ford pickup and panel trucks from 1940 through 1966. To see the colors "in the flesh," you will need to obtain a Ford brochure for your year truck. Original brochures can be found at swap meets and purchased from literature venders. The colors in the brochure won't be an exact paint match, but will help guide you in your selection if you decide to change the color scheme on your truck.

1940–1941 Car Styling Series
1940 Standard Commercial Colors

Body	Stripe
Black	Cream
Vermillion Red	Cream
Lyon Blue	Cream
Cloudmist Gray	Cream
Yosemite Green	Cream
Folkmist Gray	Red
Mandarin Maroon	Red

1941 Standard Commercial Colors

Body	Stripe
Black	Cream
Vermillion Red	Cream
Cayuga Blue	Cream
Lockhaven Green	Cream
Mayfair Maroon	Red
Palisade Gray	Red
Harbor Gray	Red

1942–1947 War-Era Series
1942 Standard Commercial Colors

Body	Stripe	Wheels
Black	Cream	Black
Vermillion Red	Cream	
Newcastle Gray	Cream	
Fathom Blue	Cream	
Moselle Maroon	Cream	
Niles Blue Green	Cream	
Florentine Blue	Cream	

Civilian truck production was suspended for 1943 and 1944.

1944
Phoebe Gray Metallic*†

1945

Body	Stripe	Wheel
Phoebe Gray*	Tacoma Cream	Black
Village Green*	Tacoma Cream	

1946–1947 Standard Commercial Colors

Body	Stripe	Wheels	Front Bumper
Greenfield Green*	Tacoma Cream	Black	Black
Medium-Luster Black	Tacoma Cream		
Vermillion Red	Tacoma Cream		

*Mercury color
†In 1944, the U.S. government gave Ford permission to manufacture a few 1-1/2-ton chassis and cabs to help alleviate the truck shortage. All were painted Phoebe Gray Metallic, a Mercury color.

1948–1952 F-1 Series
1948 Standard Commercial Colors

Body	Wheels	Grille	Grille Recess
Birch Gray	Black	Chrome bars	Argent Tan
Arabian Green			
Chrome Yellow			
Black			
Vermillion			

1949 Standard Commercial Colors

Body	Wheels	Grille	Grille Recess
Birch Gray	Body color	Argent Silver	Aluminum
Meadow Green			
Chrome Yellow			
Black			
Vermillion			

1950 Standard Commercial Colors

Body	Wheels	Grille	Grille Recess
Sheridan Blue	Body color	Argent Silver	Aluminum
Meadow Green			
Silverstone Gray			
Raven Black			
Vermillion			
Palisade Green			

1951 Standard Commercial Colors

Body	Wheels	Grille	Grille Recess
Sheridan Blue	Body color	Argent Silver	Aluminum
Sea Island Green			
Silverstone Gray			
Raven Black			
Vermillion			
Meadow Green			
Alpine Blue			

1952 Standard Commercial Colors

Body	Wheels	Grille
Sheridan Blue	Body color	Argent Silver
Glen Mist Green		
Woodsmoke Gray		
Raven Black		
Vermillion		
Meadow Green		
Sandpiper Tan		

Notes: F-1 series color combinations

Panel models: taillight and filler cap—bright finish; running boards, gas tank filler neck, and rear parking lamp—body color

Other models: frame, bumper, running boards, rearview mirror, gas tank filler neck and cap, and taillight—black; fenders and body, including interior metal surfaces before trimming—body color

1953–1956 Classic F-100 Series
1953 Standard Commercial Colors

Body	Wheels	Grille
Raven Black	Body color	Cream
Sheridan Blue		
Glacier Blue		
Woodsmoke Gray		
Seafoam Green		
Vermillion		

1954 Standard Commercial Colors

Body	Wheels	Grille
Raven Black	Body color	Cream
Sheridan Blue		
Glacier Blue		
Dovetone Gray		
Sea Haze Green		
Meadow Green		
Vermillion		

1955 Standard Commercial Colors

Body	Wheels	Grille
Raven Black	Snowshoe White	Snowshoe White
Aquatone Blue		
Banner Blue		
Sea Sprite Green		
Meadow Green		
Goldenrod Yellow		
Vermillion		
Two-tone option*		

1956 Standard Commercial Colors

Body	Wheels	Grille
Raven Black	Snowshoe White	Snowshoe White
Diamond Blue		
Nocturne Blue		
Meadowmist Green		
Meadow Green		
Platinum Gray		
Goldenrod Yellow		
Vermillion		
Two-tone option†		

Notes: Color combinations
Body color applied to the running board
*The two-tone option available on Custom Cab models consisted of a Snowshoe White roof and upper back panel over any body color.
†The two-tone option consisted of a Colonial White roof and upper back panel over any body color.

1957–1960 Styleside Series

1957 Standard Commercial Body Colors
Raven Black
Starmist Blue
Midnight Blue
Meadow Green
Willow Green
Woodsmoke Gray
Inca Gold
Vermillion
Colonial White
Prime

1958 Standard Commercial Body Colors
Raven Black
Azure Blue
Midnight Blue
Meadow Green
Seaspray Green
Gunmetal Gray
Goldenrod Yellow
Vermillion
Colonial White
Prime

1959 Standard Commercial Body Colors
Raven Black
Academy Blue
Wedgewood Blue
Meadow Green
April Green
Indian Turquoise
Goldenrod Yellow
Vermillion
Colonial White
Prime
Two-tone option*

1960 Standard Commercial Body Colors
Raven Black
Academy Blue
Skymist Blue
Adriatic Green
Holly Green
Caribbean Turquoise
Goldenrod Yellow
Monte Carlo Red
Corinthian White
Prime

Notes: Color combinations
When prime was selected, it was applied to the hood, fenders, cowl, body, driver's compartment interior, inboard step, and gas filler cap.

On trucks not fitted with Custom Cab brightmetal trim or extracost chrome, the overall color scheme consisted of the wheels, grille, bumpers, headlight hoods, and parking light rims painted white; frame, springs, axle, fuel tank, door vent frame, and glass channel painted black; and load compartment interior painted gray. On panel deliveries, the interior was painted tan.

*The two-tone color schemes consisted of any body color plus white. The contrasting white now covered the upper body and hood, cab pillars, and a sculpted panel that arched above the doors and crossed the roof behind the windshield header.

1961–1966 F-100 Series

1961 Standard Commercial Body Colors
Raven Black
Academy Blue

Starlight Blue
Holly Green
Mint Green
Caribbean Turquoise
Goldenrod Yellow
Monte Carlo Red
Corinthian White

1962 Standard Commercial Body Colors
Raven Black
Academy Blue
Baffin Blue
Holly Green
Sandshell Beige
Caribbean Turquoise
Goldenrod Yellow
Rangoon Red
Corinthian White

1963 Standard Commercial Body Colors
Raven Black
Academy Blue
Light Blue
Holly Green
Light Green
Turquoise
Goldenrod Yellow
Chrome Yellow
Rangoon Red
Sandshell Beige
Corinthian White
White

1964 Standard Commercial Body Colors
Raven Black
Academy Blue
Starlight Blue
Holly Green
Mint Green
Caribbean Turquoise
Goldenrod Yellow
Chrome Yellow
Rangoon Red
Navajo Beige
Wimbledon White*
Pure White
Bengal Tan

1965 Standard Commercial Body Colors
Raven Black
Marlin Blue
Arcadian Blue
Holly Green
Caribbean Turquoise
Springtime Yellow
Chrome Yellow
Rangoon Red
Navajo Beige
Wimbledon White†
Pure White

1966 Standard Commercial Body Colors
Raven Black
Marlin Blue
Arcadian Blue
Holly Green
Caribbean Turquoise
Springtime Yellow
Chrome Yellow
Rangoon Red
Sahara Beige
Wimbledon White†
Pure White

Note: The two-tone combinations consisted of any standard color plus Corinthian White (F-100 and F-250 series only).

　*Used with two-tone

　†The two-tone combinations consisted of Wimbledon White on the roof and around the cab back panel above the beltline, except with Special White or Chrome Yellow.

Index

Air cleaners, 74-75

Box floorboards, 196
Box restoration, 191-200
Brake line replacement, 110-112, 114
Brake rebuild, 105-107
Brake removal, 102-105
Brakes, 59-60, 102-115
Brightmetal polishing, 207-208
Brightmetal replating, 201-206

Carburetors, 68-74
Chemical paint stripping, 50-51

Degreasing, 44-45
Dent repair, 116-118

Electrical systems, 12volt, 162
Engine detailing, 88
Engine rebuild, 81-88
Exhaust manifolds, 88

Fiberglass replacement body parts, 131-133

Generators, 75-79
Glass, door and vent pane, 189-190

Horn wiring, 158

Instrument clusters, 176-178
Interior painting, 174-175
Interior repair and restoration, 166-178
Interior trim installation, 175-176

Kingpins, 60-63

Metal shrinking, 128-131

Overdrive, 93-101

Painting, finish coat, 142-145
Painting, primer coat, 135-140
Parts accessibility, 28-29
Pinstriping, 145
Plastic filler, 125-126
Plastic parts polishing, 209
Plastic parts replating, 206-207

Rebuilding, pickup truck, 19-20
Restoration, costs of, 23-27
Restoration, standards of, 23
Restoring, pickup truck, 20-23
Rubber parts, 179-189

Rust repair, 118-119
Rust/derusting, 52-53

Sandblasting, 45-50
Seat cushion recovering, 167-174
Shop needs, 33-35
Springs, 57, 63
Starters, 75, 79-80
Steering, 63-65
Storage, 210-216
Straight axle front ends, 54-66

Tie rods, 58, 63
Tool selection, 32-33
Tools, basic, 35-37
Tools, specialty, 37-41, 67-68
Transmissions, automatic, 89-90
Transmissions, manual, 90-93
Turn signals, 157-158
Twin I-Beam IFS, 66

Welding, gas, 38-39
Welding, patch panel, 123-124
Welding, wire, 119-121
Wiring, replacement, 150-165
Wiring, tools, 150-161